Howard Douglas

A Postscript to the Section on Iron Defenses

Howard Douglas

A Postscript to the Section on Iron Defenses

ISBN/EAN: 9783337158019

Printed in Europe, USA, Canada, Australia, Japan

Cover: Foto ©ninafisch / pixelio.de

More available books at **www.hansebooks.com**

A POSTSCRIPT

IRON DEFENCES,

CONTAINED IN

THE FIFTH EDITION OF 'NAVAL GUNNERY'

IN ANSWER TO THE ERRONEOUS PRINCIPLES SET FORTH BY THE REVIEWER
IN 'THE QUARTERLY REVIEW' FOR OCTOBER, 1860.

By GENERAL SIR HOWARD DOUGLAS, Bart.,

G.C.B.; G.C.M.G.; D.C.L.; F.R.S.

SECOND EDITION, WITH ADDITIONS.

LONDON:
JOHN MURRAY, ALBEMARLE STREET.
1861.

A POSTSCRIPT,

&c. ` *&c.*

In the article entitled 'Iron Sides and Wooden Walls,' in the
'Quarterly Review' for October, 1860, the Reviewer states that
"it is difficult to see by what process of reasoning Sir Howard
Douglas had arrived at the conclusions stated in the fifth edition
of the 'Naval Gunnery.'"

The article professes to be a review of the works whose titles
are prefixed, but the writer only alludes to the one and takes no
notice of the other. He neither states the facts that have been
established, nor the results obtained from the experiments
recorded in Section XII. of the 'Naval Gunnery,' nor the
question which I proposed to examine for the purpose of bringing
the whole subject of iron defences broadly before the country for
fair and impartial discussion.

I am well aware that it has become a common practice for
writers in Reviews to place at the head of their articles the titles
of books which are not even mentioned in the body of the article,
which is therefore rather an essay on the theme to which the
books relate—a practice to which a celebrated essayist, the late
lamented Lord Macaulay, was much given, but which I think
has a prejudicial effect upon literature. And if, in compliance
with this very questionable usage, the so-called Reviewer had
omitted to mention the 'Naval Gunnery,' as he did with respect to the 'Fortification,' I should not have taken any um-
brage, nor have noticed the omission; but I put it to the reader
whether the mention made of the 'Gunnery' does not imply
either a doubt of my reasoning power or an impugnment of the
integrity of my conclusions.

Before, however, I explain the process by which I arrived at
my conclusions, I must further remark that, in acknowledging
the courteous and complimentary terms in which the writer has
been pleased to speak of the regard due to my opinions on
matters of this description, his words seem to imply that I have

some official connection with the Admiralty. I have none whatever, excepting relations of great respect and regard for the noble Duke the First Lord, and for my gallant and distinguished friends the members of the naval administration. I hold no office under the Government—I derive no advantage from my labours relating to the naval and military defence of the country. Her Majesty's Government were pleased to request my services and ask my opinion upon the important subject of our defences; and, though far advanced in my eighty-fifth year, I have given all my energies to the study of the subject, and have imparted my views to the Government without receiving or desiring any other recompense than the consciousness of devoting, it may be, my last intellectual and mental exertions to securing the safety of the country. .

It has become common of late with writers in the public press, some of whom are perhaps not fully informed as to the facts and science of the case, to use sneers against those who venture to oppose what may appear to them rash innovations; but surely it is very unfair to try thus to prejudice the public mind against the opinions of men who certainly have the advantage of long experience, by declaring them to be set against all improvements, wedded to old fashions, prejudices, and worn-out practices—denunciations which should not at least be pronounced until the changes in question are proved beyond all doubt to be in the right direction.

The study of my life, for the last fifty years up to the present time, has been devoted to the reform and improvement of the naval and military services of the country. I have ever endeavoured to abolish effete, old-fashioned guns, systems, and practices of war—to improve the weapons and science of war on the flood and on the field—to watch, study, meet, and promote the changes which modern science has originated, and have never obstructed the improvements which modern science has produced. I was the first to call the attention of the Admiralty and the country to the unsatisfactory state of naval armament and naval gunnery, and to propose remedies for both; and I was the first to awaken the country to the great changes in naval warfare which the introduction of steam propulsion must produce. Whilst thus endeavouring to promote what appeared to me to be real improvements, I have endeavoured to be a drag on what I thought rash, hasty, extravagant propositions which have been brought forward within the last twenty years; upon which vast sums of public money have been wasted, but which would have been saved to the country had my advice upon the several projects which I am now to enumerate been adopted.

I arrived in England from Corfu late in the autumn of 1841, and on the 9th of February, 1842, took my seat in the House of Commons as member for Liverpool. In the course of that session I was consulted by the late Sir Robert Peel, as to the use and efficiency of a certain half-dozen iron frigates, two of which were finished and four constructing by contract. I stated in reply that vessels wholly constructed of iron were utterly unfit for all the purposes of war, whether armed or as transports for the conveyance of troops. I stated that a shot striking with great velocity would drive into the ship numerous splinters of the disk struck—that shot striking with reduced velocity, as when fired from a distance, would make large jagged holes that could not be plugged from the inside—that shot striking a rivet or rivets, would make a large breach in the side of the ship—that the shot might break on impact, and its fragments, together with those of the plate, would drive into the ship a mass of splinters consisting of pieces of the shot, bolts, bolt-heads, nuts, and innumerable pieces of iron, which would prove far more deadly and extensive than any case-shot. I strongly urged Sir Robert Peel not to sanction the appropriation of any public money to the project, unil the efficiency of such vessels should have been fully tested by actual experiment. Some time after this, I received a note from Admiral Sir George Cockburn, acquainting me that experiments were to be made at Woolwich, which he understood I was to attend, and proposed to me to accompany him to Woolwich, in the Admiralty barge, for that purpose; which I accordingly did, and there witnessed the experiments. After the firing ceased, I picked up a bag full of fragments of iron of every description —pieces of shot, bolts, rivets, and numerous splinters of plates, and found abundant proof how forcibly these had been driven into the ship, and against the opposite side.[1] These are the very fragments described in the table at the foot of page 132, stated to have been before me when I wrote the retrospective

---

[1] 'Naval Gunnery,' 5th edit., Art. 388, p. 396.

observations, in Article 175, on the effects of shot on iron ships, when preparing my work for a new edition in 1851. I exhibited those fragments to Sir Robert Peel on my return to London, together with a verbal Report on the results. Those experiments put an end to the construction of iron vessels at that time. Vain attempts having been made to back the plates up with a composition of sawdust and caoutchouc—which made the matter rather worse, and, being highly inflammable, created a fresh danger,—it was then proposed to use the two frigates that had been constructed, as transports for the conveyance of troops. The serious objections I urged against this use of vessels constructed of iron are confirmed by the melancholy loss of the "Birkenhead."[2] I predicted the failure of the Lancaster gun,

---

[2] I maintain that had the "Birkenhead" and the "Royal Charter" been timber ships, the regiment that perished in the former, and the hundreds of passengers that perished in that most appalling catastrophe, the wreck of the latter, would have been saved.

The following remarks bearing on the risk peculiar to iron-built ships, and referring to the wrecks of the "Birkenhead" and "Royal Charter," are extracted from my 'Naval Gunnery,' 5th edit., pp. 133-5.

"In vessels made of iron the weight of the whole material is considerably less than that of vessels of the same dimensions but made of timber. With equal displacement (the weight of water displaced), therefore, the former can carry a greater weight of cargo, their capacity for stowage being greater on account of the thinness of their shell. But when an iron vessel is bilged and becomes filled with water, the superior weight of the material of which she is formed is a momentous consideration. Timber, when immersed in water, loses as much of its weight as is equal to that of the water displaced by it, and it floats. A cubic foot of oak timber has a buoyancy, when immersed in salt water, of 76 oz., and a cubic foot of fir, a buoyancy of 450 oz. ; but the excess of the weight of a cubic foot of iron over an equal volume of salt water is 6180 oz., and, with this force, the iron sinks. When, therefore, an iron ship is bilged, having lost its power of floating, the weight of the iron tends to break and destroy it, unless it be stranded on a smooth and shelving bank or beach. This was the case with the 'Great Britain,' which rested, throughout her length, upon the beach on which she grounded; but when an iron vessel strikes upon, and is perforated by a rock (as was the case with the 'Birkenhead,' which was wrecked near the Cape of Good Hope), she becomes bilged, and there she remains with deep water at her extremities; she then becomes filled, either wholly or partially, and the iron, deprived of buoyancy, exerts a prodigious force to break the vessel's back and sink the portions which are not in contact with the rock. Even if the portion which is not in such contact should be furnished with compartments which may not permit the water to enter, the difference between the power of flotation in that water-borne portion, which acts in a contrary direction to the weight of the iron, will constitute a strain which no iron vessel can resist, tending to fracture it at the section which divides the filled from the unfilled portion. Should there be no water-tight compartments, or should these not effectually act, then an iron vessel, perforated near the midship's section, with deep water under her extremities, will infallibly be destroyed by the weight of the iron at both ends acting upon the fulcrum on which the vessel rests.

"What has been said is applicable to the most melancholy case of the 'Royal Charter,' an iron screw steamer from Melbourne, which was wrecked on our

and did all in my power to dissuade the Ordnance authorities from adopting that scheme: I was not listened to, and enormous sums have been wasted on that proposition. I predicted the failure of the Dundas built-up gun; the monster mortar (p. 175 of the 5th edit. contains a diagram of the Mallet mortar); the Nasmyth wrought-iron gun; the Liverpool gun, not from much apprehension of want of strength, but on account of the impossibility of installing such a gun in any ship with which to attack Cronstadt, for which purpose the gun was made and presented to the Government. I predicted the failure of the floating batteries first constructed,[3] and of the iron frigates "Erebus" and

own coast, near Liverpool, during the fearful storm in October, 1859, when the vessel broke transversely in two places, and above 400 persons perished.

"When an iron vessel parts, all the fragments of the material of which she is composed sink, and nothing floats to save life but a few loose spars: whereas the fragments of a timber ship float, and many, perhaps most of the crew, may thus be saved.

"Applying these observations to the case of the 'Birkenhead,' and on a full consideration of the facts as taken from the report of the proceedings of the court martial on the surviving officers and crew of that ship, there cannot be the least doubt that, had she been a wooden vessel, she would have held together long enough in a sea so smooth as to permit all her hands to be saved. To this conclusion it may be objected, by persons who have not duly considered the difference of circumstances, that the 'Avenger,' though constructed of timber, went to pieces even more quickly than the 'Birkenhead,' when, going at full speed, she struck upon a reef of rocks; and, with the exception of a small boat's crew, all hands perished; but, on that occasion, the sea broke over the reef with such violence as to render it impossible that any vessel could resist it, or that any of the crew could be saved by clinging to the fragments into which it was broken."

In further confirmation of the above remarks I would refer to the account of the loss of the "Connaught," given in 'Mitchell's Steam-Shipping Journal,' which shows that she left Galway on the 25th September, and on the 6th of October sprung a leak, and sunk in deep water, the origin of the leak being therefore unattainable with certainty. The length of the "Connaught" was 370 feet, her breadth of beam 40 feet. The shells of the ship were from ⅞ to 1¼ inches thick, and frames doubled in the centre along a section of 270 feet. After examining the details at some length, the conclusion arrived at was, that the sinking of the "Connaught" must have arisen through excessive tension tearing away the plates amidships, where the heaviest were deposited—ships of such enormous lengths fall in with waves that take the bow and stern, and suspend momentarily the midship section; and that no other conjecture or solution can be given of the loss of the "Connaught" than this suspension theory.

Among the vessels wrecked in the fearful storm in the Baltic at the commencement of October last, were two iron steamers, the "Edinburgh" and the "Moscow," which, in addition to the risk incurred by their being iron-built ships, had to contend with a source of danger which rendered the wreck inevitable—the being over-burdened with top-weight. It is stated that the "Edinburgh" had a deck-load of nearly 500 bales of cotton in addition to two steam-boilers; and that the "Moscow" had an immense deck-load, chiefly cotton bales, piled up the whole length of the deck. This is a risk, however, which should be prohibited by legislative enactment.

[3] 'Naval Gunnery,' 4th edit., pp. 358-362, and App. G.

"Meteor," each of which cost 50,000*l.*, and expressed my conviction that the project of the iron-chain portcullis would be a
failure.[4]

I was, much against my will, appointed chairman of the Commission for examining and reporting upon the Warner inventions.
I need not go over that matter, it will be in remembrance of most
of my readers. This I may say, that the impression produced
by the press was that there was something new and astounding in Mr. Warner's alleged inventions which, irrespective of
their terrible efficacy in war, it behoved us to purchase and for
ever conceal for the sake of humanity. Where now are Mr.
Warner's alleged inventions, which he proposed to sell to the
Government for 400,000*l.*? Had it not been for the Commission, of which I was the chairman, and for the opposition I
made to that imposition in my seat in the House of Commons, to
which appeals were made, Mr. Warner would assuredly have succeeded in obtaining a large sum of money from the Government,
for the sale of alleged inventions that were not worth sixpence—
the one being a balloon and the other a trick. Yet never was a
public man, in the discharge of a public duty, more taunted and
assailed than myself—as "prejudiced," "antiquated," "a red-
tapist," "nothing like leather," and other such-like vulgarities.

And here I would remark that the sarcastic, and I must say un-
feeling, taunt in p. 556 of the Review might have been spared. It
could only do violence to the memories and kindred of the dead,
and the feelings of the living. The old soldiers of the Peninsula
might have been permitted to pass away in peace to their graves,

---

[4] Though not immediately connected with the question of our defences, I
may state that, as a member of the Harbour of Refuge Commission, I protested
formally against the mode of construction adopted by a large majority of the
Commission for forming a harbour of refuge in Dover Bay, on the engineering
question of building the piers and breakwater of upright walls of blocks of
granite, and of concrete, by the slow process of using the diving-bell. But the
main grounds of my protest were expense and time of execution. The Commission estimated the work at 2½ millions, and the time of execution five or
six years, or some such comparatively short period. I protested that such
would be the difficulty in building an upright wall in 7¼ fathoms water,—
such the disturbances and dilapidations occasioned by storms and the per-
cussive force of the waves, that the work would cost nearer 10 millions than 3;
and, as to the time of execution, that the youngest human being then living
would never see it finished. And now, in 1860, having watched year by year
the progress of the work, the rate at which the pier has been extended per
annum, and the enormous sums that have already been expended upon the
work, I assert that it cannot be completed, according to this mode of construction, for sixty or seventy years from the present time.

without being upbraided, in the pages of a national chronicle of world-wide reputation and circulation, for the respect and consideration in which they naturally held the weapons—however imperfect when compared with those now in use—with which they first hurled back the full tide of French conquest, in that most righteous and retributive war which England undertook in the Peninsula, for the independence of nations inhabiting that part of the world, and following up that tide in its ebb from victory to victory with the bayonet in the reins of the discomfited foe; ultimately crushing despotism, emancipating Europe, and obtaining an enduring peace on the glorious plains of Waterloo.

At an early period of my long-continued labours for the improvement of gunnery, I applied myself to the rifle principle, and in 1826, second edition, suggested the advantage of that principle, in treating of oblong shot double the weight of spherical shot which were then in use, in close action, from the large fracture which they made in the sides of ships, but which shot could not be used excepting in very close action, from the great irregularity in their flight, tumbling over and over, and not penetrating but striking with their side, unless made to rotate by being fired from a rifle gun; and this observation was the origin of attempts, soon after made, to effect this great object. I propounded the rifle principle for muskets in 1851, founded upon Newton's celebrated theorem on the solid of least resistance, and deduced from it the form which an elongated rifle-shot the most likely to satisfy the conditions of that theorem should have. It was a step in the right direction, to consult the principal rifle-makers in London respecting the construction of a rifle-musket, to form a model on which to proceed in the manufacture of the best rifles for the service of the army; and thus the Enfield rifle originated. But I never thought that rifle the best that could be produced, but that a better would be forthcoming on the Whitworth principle; and a better is now before the public, having been tried at Hythe, and reported on by Major-General Hay ('Naval Gunnery' 5th edit., pp. 214, 569). The superiority of the Whitworth rifle is universally admitted; but, having gone so far as to be unable to recede, we are now precluded from availing ourselves of that superiority, without a vast sacrifice of money invested in the establishment for manufacturing the Enfield rifle, with which the British army is now provided.

This is all very well as a measure of economy, but how will it be elsewhere? The French never adopted the Minié rifle, although Colonel Minié is instructor of the School of Musketry at Vincennes. The French have been lying by on the question of rifles, satisfied in the meantime with the "carabines à tiges," till the problem of rifles should be worked out. Profiting by the experience of others, it may be that they will produce a better rifle-musket than ours. Mr. Whitworth is negociating with the Emperor Napoleon III. for the sale of the French patent; and no doubt it will be purchased, if we are so supine as not to purchase immediately the whole patent right, French and English, from Mr. Whitworth, to prevent a superior weapon, the invention of an Englishman, from being used against us.[5] I thought we went too fast and too far in believing that guns for firing elongated rifle-projectiles would supersede the old smooth-bored artillery in all descriptions of service, and form exclusively the broadside armament of all our ships; for I much dreaded the introduction of breech-loading guns into the naval service, until the safety to the users of that description of gun should have been fully tested under all circumstances, resembling as much as possible those of real service in protracted and rapid firing, as in battle. Much of this I doubted,

---

[5] The 'Times,' in allusion to the recent "Tir," stated that, the emperor becoming aware of Mr. Whitworth's presence, appointed an interview with him. He had several of Mr. Whitworth's rifles in his possession, and objected that they had been found to foul. Mr. Whitworth replied that with fair treatment this was not possible, and agreed to compete at Vincennes with any rifle that could be brought against his own. The range was marked out at 500, 700, and 1000 metres, at Vincennes The result of the trial is thus described:—"The superiority of the Whitworth rifle was so manifest that at the 700 metres' range the French marksman retired from the contest completely discomfited. On Wednesday morning it was intimated to Mr. W. that the experiment was very satisfactory; that his majesty wished a number of rifles to be made for him; that he would send an officer to Southport to see the cannon tested as soon as arrangements could be made for that purpose; and that, provided no objection were presented by the nature of the ammunition used, he would at once negotiate for the purchase of the French patent, so as to make the invention available for the service. As this might lead to inferences injurious to Mr. Whitworth's patriotism, it is only fair to state that early in 1857 he clearly demonstrated at Hythe, in the presence of Lord Panmure, then minister of war, the great superiority of his rifle over the Enfield. During the long period which has since elapsed no serious steps have been taken by the English government to secure for the army or navy either the cannon or the rifle. Napoleon III., having won Solferino by his new artillery, is not the man to neglect a chance for excelling the Enfield and Armstrong, and, if he accomplish this by jumping at a mechanical invention, which our War Department, after in a great measure paying for it, has neglected to utilize, his triumph will be complete."

as I have stated in Article 243, pp. 231-7, 5th edit.; nor did I believe that elongated shot are as well adapted to siege operations for battering in breech, nor as well adapted to perform the office of the mortar as spherical shells, nor capable of that most important description of ricochet firing by which Vauban gave the superiority to the attack of fortified places, the balance having been previously in favour of the defence over the attack. Of this disability of elongated projectiles I have collected many proofs. It suffices now to put that doubt beyond all question, by referring to the second round fired at the Eastbourne tower from a 7-inch howitzer, from a battery distant 1032 yards, which struck the shingle about 65 yards short, turned over, and ricocheted over the tower. Many similar facts have been established in firing the 12-pounder Armstrong field-gun point-blank, or at low angles: sometimes the projectile ricochets at a great angle, rises to a great height, grazing next at a great distance, and deviating greatly from the line in which the gun had been laid.

In the late experiments against the tower at Eastbourne, some other important incidents occurred which show defects in the Armstrong shells and fuzes which merit very serious consideration, and should be made the subject of extensive trials, as recommended in 'Naval Gunnery,' p. 235, 5th edit. One 40-pounder shell and five 6-inch shells burst prematurely a little way in front of the gun, which can only be attributed either to the interior of the shell having been left in too rough a state and a consequent ignition of the bursting-charge by its friction in the shell, or to the shell being of insufficient strength, or to defects in the fuze. Of the percussion-fuzes employed, many exploded with comparatively little effect in grazing, and in passing through the débris of the tower few, if any, penetrated the brickwork before exploding.[6]

The questions which I proposed to examine in my 'Naval Gunnery' were as follows:—

1st. Whether ships constructed wholly or nearly so of iron, are fit for any of the purposes or contingencies of war.

---

[6] In consequence of the number of Armstrong shells that had failed to explode, it became necessary to use great precaution in removing the débris of the towers, and a notice was printed and posted by the authorities to guard against any risk being encountered; but, notwithstanding this, a workman foolishly sacrificed his life by applying a light to one of the unexploded missiles.

2nd. Whether ships constructed of timber can, with due regard to the conditions on which flotation, stability, manageability, and safety depend, be covered with plates of ponderous, rigid, and brittle material, of adequate thickness to resist the penetration and withstand the impacts of the powerful ordnance still in use, and the more potent guns and precise projectiles which are now being introduced.

I now proceed to explain the process by which I arrived at the conclusions stated in the 'Naval Gunnery,' and repeated in p. 15 of this Postscript.

At vast labour and no small expense I collected accounts of all the experiments made to test the resistance of metal plates

---

[7] DESCRIPTION OF PLATES tried in the Experiments at Portsmouth and Shoeburyness, with the Names of the Makers.

Date of Report.	Description of Plate.	Maker's Name.
21 Aug. 1858 . .	Common wrought-iron plates, 4 inches thick	
,, . .	Common iron plates, 6 in number, rivetted together, making the total thickness 4 inches . .	
,, . .	Longitudinal bars of iron, 4 inches thick and 3¼ inches wide, having African oak between them of the same dimensions . .	Messrs. Shortridge, Howell, and Jessop.
,, . .	Homogeneous metal plates, 4 inches, 3 inches, and 1¾ inches thick, from 1 foot 6 inches to 2 feet 6 inches in breadth . .	Do. do.
,, . .	2-inch homogeneous metal, covered with 3-inch African oak plank .	Do. do.
,, . .	4-inch and 3-inch steel plates .	Messrs. Westwood, Baillie, and Campbell.
,, . .	2½-inch and 2-inch steel plates .	Messrs. Naylor, Vickers, and Co.
26 Oct. 1858 . .	Iron sides of "Erebus" . . .	
10 Nov. 1858 . .	Iron side of "Meteor" . . .	
19 May, 1859 . .	4½-inch wrought-iron plates, 3 feet broad, 9 feet long, 3 in number, two upper ones rolled and the lower one forged. Under the plates was 3-inch teak bedded on ⅝ wrought-iron plate, which was again attached to vertical feather-plates ₇⁄₁₆-inch thick and 18 inches apart, forming cells 18 inches wide and 18 inches deep. These plates were then attached to the inside iron skin of the target, 1 inch thick ; the whole thus representing the section of the hull of a ship, ribs and all complete . .	Messrs. Palmer and Co.

and other materials since 1838, in France, the United States, and in Great Britain, and more particularly of those made at Portsmouth, Woolwich, Shoeburyness, and Chatham, since 1845. These experiments were made with ordnance of every description, and projectiles of every nature and form, at distances varying from 200 to 1500 yards and upwards, with every variety of charge, against masses of timber, earthen defences, masonry defences, and slabs of cast and wrought iron—against metal plates of every description—steel plates of 2, 3, and $3\frac{1}{2}$ inches, of the qualities known as red, white, &c.—against plates of wrought iron, plates of iron wire welded together, against plates of homogeneous iron, and puddled iron.

The plates [7] were furnished by the most renowned iron

DESCRIPTION OF PLATES, &c.—*continued.*

Date of Report.	Description of Plate.	Maker's Name.
19 Aug. 1859 . .	Hammered puddle-steel, 6 feet by 3 feet, 3 inches thick . . .	Messrs. Mare and Co.
,, . .	Hammered puddle-steel, 6 feet by 3 feet, $2\frac{1}{2}$ inches thick . . .	Do.
,, . .	Same material, but made by another process, 6 feet by 2 feet, 3 inches thick	
,, . .	Same material, but rolled, 6 feet by 2 feet, $2\frac{1}{2}$ inches thick . .	
,, . .	Plates $4\frac{1}{2}$ inches thick annealed, 6 feet by 3 feet	Thames Iron Works.
,, . .	Plates $4\frac{1}{2}$ inches thick not annealed, 6 feet by 3 feet . . .	Do.
,, . .	Plates $4\frac{1}{2}$ inches thick, rolled iron, 6 feet by 3 feet	Messrs. Palmer and Co.
,, . .	Do. do. 9 feet by 3 feet	Do.
,. . .	Plates $4\frac{1}{2}$ inches thick, 9 feet by 3 feet	Portsmouth Dock Yard.
12 Oct. 1859 . .	Rolled iron plates, 4 inches, $3\frac{1}{2}$ and 3 inches thick	Messrs. Palmer and Co.
,, . .	Steel plates, 3 inches thick . .	Sir Charles Fox.
22 Feb. 1860 . .	Wrought-iron plate, $4\frac{1}{2}$ inches thick, 6 feet by 3 feet . . .	Swedish Steel-iron Company.
1 March, 1860, at Shoeburyness .	Rolled scrap-iron plate, $4\frac{1}{2}$ inches thick, secured to a butt, representing the section of a frigate .	Messrs. Palmer and Co.
20 March, 1860. .	Wrought-iron plate, $4\frac{1}{2}$ inches thick, 6 feet by 3 feet . . .	St. Helen's Iron Works.
,, . .	Wrought-iron, $4\frac{1}{2}$ inches thick .	Darlington Forge Company.
28 March, 1860. .	Iron plates, $4\frac{1}{2}$ inches thick . .	Messrs. Westwood, Baillie, and Co.
11 May, 1860 . .	Armour-plate for "Warrior" . .	
24 May, 1860 . .	Armour-plates for "Resistance" .	Messrs. Westwood, Baillie, and Co.
6 June, 1860 . .	Armour-plates for "Black Prince"	Messrs. Napier.
2 July, 1860 . .	Plate from side of "Trusty" . .	

masters of England and Scotland, who were invited to send for trial metal plates of various qualities and thicknesses, for the purpose of trying which should be found to combine the greatest power of resistance with the least weight.

I attended many of these experiments, and vouch for the accuracy and authenticity of the results which I have recorded in Section XII. of the 'Naval Gunnery.'

In examining and considering the facts and results of these most valuable experiments, I could not but feel that the country is greatly indebted to the several naval administrations, by authority of which these trials were instituted, in furnishing, at the public expense, facts which no individual could otherwise have obtained, and from which only sound and safe conclusions could be obtained in the several matters on the points in question; and I could not but admire the wisdom, discretion, and steadiness of the naval administrations severally, in not being driven out of the course they had so wisely taken, by any clamour of the press; for though in matters of opinion, policy, &c., it may be a competent tribunal, yet in matters of fact and exact science, the organs and exponents of public opinion are really not qualified to investigate and decide. Sensible of the vast advantages of these well-established facts and results, in enabling me to deduce right conclusions from a careful examination of those data, I resolved to publish them in my work, as the only sure grounds on which the whole question of iron defences might be brought broadly before the country, for that full, impartial, and calm investigation and discussion, which it is of the utmost importance to the country that questions of fact should undergo. For this I sent copies of the work to the 'Times' and other leading journals on the 2nd of August last; and, if I may judge from some articles that have appeared in various influential journals, and from the altered tone of those who had, I think, proceeded upon errors of fact, I may hope that my efforts in promoting calm discussion upon this question have not been unsuccessful, and that the question is more likely to be calmly and soundly discussed, with clear regard to the facts that I have endeavoured to lay before the country. As an instance, I may quote the following admissions, quite consonant with the statements of the 'Naval Gunnery,' which occur in the course of an article in the 'Times' of October 26th, advocating the application of heavy armour to ships:—"We do not yet know how far this ponderous armour may be compatible with the sailing or steaming qualities of the vessel." "It has been reported that the 'Gloire' has proved a failure as a sea-going vessel." She "was never intended to go to sea," but "designed for home service only." The "masts may be shot overboard," and thus "the screw fouled, and the vessel rendered helpless." "Iron-cased ships will be calculated for home service only"; "our foreign and colonial service must be performed by old-fashioned vessels."

After a close and careful examination of all the facts established, and the results obtained from these experiments, I came to the following conclusions:—

1st. That ships formed wholly or nearly so of iron, are utterly unfit for all the purposes and contingencies of war, whether as fighting-ships or as transports for troops: 68-pounder solid shot would pass through and through the "Great Eastern" with tremendous effect, and the perforation in the outer shell could not be plugged; she is an awful roller, and has never attained anything like calculated speed.

2nd. That thin plates of wrought iron even $\frac{5}{8}$ of an inch thick are proof against shells or hollow shot in an unbroken state, but that the fragments of the shot or shell pass through the plates and produce an effect perhaps more formidable than any shell.

3rd. That being proof against shells will avail little (Art 441) unless the vessels are likewise proof against solid shot; for shells would, of course, not be fired against ships proof against them, whereas the destructive effects produced by fragments of shot and of plates, and the great damage done to the scantling of the ship by solid shot appear more like the result of a shell than of a shot.

4th. That elongated projectiles produce greater effect than spherical projectiles of the same weight at long than at short ranges, on account of the rifled elongated projectiles—the resistance to which is a minimum—retaining more of their initial velocity than spherical projectiles at the same distance.

5th. That the thickness of plates required to resist shot fired from the heaviest nature of guns must not be less than $4\frac{1}{2}$ inches.

6th. That to secure the resistance of the plates and the impenetrability of the sides of a ship, it is indispensable that the plates be strongly backed by masses of the strongest and most resisting timber, as, in all the cases to which reference has just been made, it appears that the plates are easily broken when the support is removed from behind them, by the crushing, fracturing, and damaging effects of the impacts of the shot.

7th. That no iron-cased vessel with upright or nearly upright sides, impenetrable to or invulnerable by heavy shot, and fulfilling all the conditions on which flotation, stability, and manageability depend, has yet been produced.[8]

8th. That the project of strengthening the faces of land-defences of masonry with plates of iron has not succeeded; and that the very worst combination that can be made of materials for defensive purposes is that of iron and stone.

My opinion, and the grounds on which I formed it, that vessels made wholly of iron are unfit for any purposes of war, are thus stated in my 'Naval Gunnery,' 5th edit., Art. 176:—

" It is generally believed that iron vessels, however convenient and advantageous in other respects, are utterly unfit for purposes of war. This opinion has been confirmed by the decision of a mixed committee of officers of the Naval Artillery and Engineers. This Committee was appointed under the authority of the Admiralty and Board of Ordnance, in order to consider how far it might be possible to carry into effect a plan for arming the Contract Mail Packet Steamers, and to report whether or not the terms of their contract have been observed by the several companies as regards the adaptation of their vessels for war purposes. From this category the Committee entirely and unanimously reject iron vessels. Exclusive of these, they found that of the fifty-three vessels belonging to the Peninsular and Oriental and the West India Mail Companies, eight only (of wood) were capable of being effectually fitted to receive an armament especially directed to the object of defence ; though some of the other ships might be fitted to serve as armed packets or troop-ships. In almost every case it was considered

---

[8] The propositions for forming iron-sided ships with angulated sides will be considered further on.

impracticable to have a pivot-gun either forward or abaft, on account of the sharp form of the bow and the great rake of the stern."

The system of iron-cased vessels is not—as stated pp. 559, 560, of the Article in the 'Quarterly Review'—a novelty, nor the invention of the Emperor Napoleon III.: it was the discovery of Colonel Paixhans, the well-known inventor of the Paixhans canon-obusier gun; and proposed by him nearly forty years ago. The proposition was referred to the Comité Consultatif de la Marine and to the Institute of France, and reported on unfavourably; on the ground that no vessel, even a line-of-battle ship cut down to the lower deck, could carry such an immense and very expensive weight of armour sufficiently thick to render it invulnerable. It was therefore abandoned, but has been reproduced by Napoleon III.: first in the form of the floating batteries originally constructed in France and copied by us in the late Russian war—all of which were failures; and subsequently in the form of "La Gloire" *frégate blindée*, covered, it has been said, with plates of newly-discovered qualities, much lighter and less penetrable than wrought iron. That ship is not a success, and the alleged discovery is fiction, as I shall hereafter show.

In the United States, after elaborate experiments with iron plates of different thicknesses and qualities, the same conclusion with respect to the thickness of the plates (viz. 6 inches) was arrived at, and the project was abandoned, as it had previously been in France.

By a recent communication from the United States, I learn that twenty steam-vessels have been added to the navy of the United States, since the commencement of the present administration—thirteen by construction, and seven (named in the Report) by purchase. Those built are—the "Lancaster," 'Pensacola," "Brooklyn," "Hartford," and "Richmond," having steam power as auxiliary to sails, being armed with heavy 9, 10, and 11 inch Dahlgren shell-guns, and having a speed of 12 statute miles an hour at sea under steam alone; the seven new screw sloops of war, "Mohican," "Narragansett," "Iroquois," "Wyoming," "Pawnee," "Dacotah," and "Seminole;" and the side-wheel steamer "Saginaw." The contractors for the machinery have

B

guaranteed the speed—for the "Pawnee," 16 statute miles an hour, for the "Dacotah," 15 miles, both under heavy penalties; the other five, eighty revolutions of the propeller in a minute, a velocity which in the "Wyoming," the only one yet tried, has given a velocity of 14 miles an hour, without the use of sails.

Although the naval force of the United States has been thus increased by the addition of twenty screw steam-ships, a still further increase of the navy has been recommended in the report of the Secretary of the Navy. But no iron-cased vessel is built or being built, nor even alluded to in any way in that very able Report on the increase of the navy, nor in the miscellaneous observations and contemplated changes. The experiments tried in the United States, on the project of endeavouring to render ships proof against shot, by covering their sides with iron plates of adequate thickness, having proved that nothing less than 6 inches will suffice; the proposition for strengthening masonry works and casemated embrasures by blocks and slabs of iron having failed, ('Naval Gunnery,' Section XII.); and a commission of naval officers having deprecated the proposition to apply naval resources to coast defences in the form of floating batteries, the Government of the United States are satisfied as to the futility of all such expedients, and probably consider all those questions settled in the negative.[*]

These conclusions may safely be added to those given in p. 15, namely, that though plates $4\frac{1}{2}$ inches thick, if well backed up with masses of solid timber, may for a time resist the penetration of shot fired at considerable distances, yet that a vessel

---

[*] In the course of an article in the 'Times' of October 31st, 1860, on the Report of the "United States' Naval Board of Inquiry," the following statements are made :—The Americans are looking to their national defences, and surveying their navy. Not a syllable, however, is said of *casing ships of war in iron :* not a hint given of substituting iron for wood. They have generally anticipated us in the adoption of new principles : they had a steam-ram on the stocks before Admiral Sartorius introduced the invention to the public. The scheme of casing ships of war in iron was undoubtedly brought before the United States' authorities some years since; but was not approved— whether the experiments were unsatisfactory, or that they trusted to the superior penetrating power of their Dahlgren guns, does not appear. They have never taken up the new theory and do not appear to be embracing it now.

Many points are still in dispute, viz. :—That the "Warrior" is very complicated ; that her sides are thought too straight : the question being not only as to impact of shot, but the proper adjustment of weight ; as the sides being sloped inwards would bring the enormous weight of guns and armour more to the

cannot be rendered invulnerable to shot by iron plates less than 6 inches thick, a weight which no vessel can carry without great sacrifice of speed, great instability, and incapability for open sea and ocean service.

The experiments stated in the Section on "Iron Defences" having proved that no perfectly shot-proof vessel with upright sides has yet been produced, a patent has been obtained by Mr. Josiah Jones (1st November 1859, No. 2191), now before me, for constructing vessels whose sides above the water-line slope inwards at an angle of 45°, so that a horizontal shot fired at a vessel's side, by striking obliquely, may not penetrate, though if upright the shot would pass through.

Mr. Jones's patent consists in applying steel and iron plates, 3½ and 4½ inches thick respectively, in combination with ribs of framing of timber, to a ship constructed with inclined sides; the ship being formed with an angular bend or projection in an outward direction at the line of flotation, so that a shot will glance off either upwards or downwards, according as she may be struck above or below the line of flotation. In examining new and untried projects, I invariably let the inventor state his own case. Thus, the above account is taken from Mr. Josiah Jones's specification in his provisional patent, and the drawing of the body of the ship reduced from that which is filed with Mr. Jones's patent. (See next page.)

Experiments were made at Portsmouth in August, 1860, to ascertain the effect of heavy shot on a butt covered with 4½-in. iron plates when placed obliquely to the blows. The butt to which the plates were attached, which was intended to repre-

---

centre, and thus improve the motion, while too great a slope, as in Jones's model, would leave the vessel to be washed over by the waves.

That it is doubtful whether the best model has been adopted, and more than doubtful whether the best means have been taken to arrive at the truth.

That no experiments have been directed to the solution of the difficult problem of *combining an impervious armour with the sea-going qualities* of the vessel so protected. No iron-cased ship having been so much as floated *to show how such a ship would sail or steam.*

That we are still working in the dark; that the *matter being uncertain*, we must feel our way: a single vessel should have been experimentized with at once by testing her sea-going qualities, instead of venturing on what may prove five failures.

That the new system being in its "infancy" may lead us to hopeless and endless extravagance, unless timely precautions are taken.

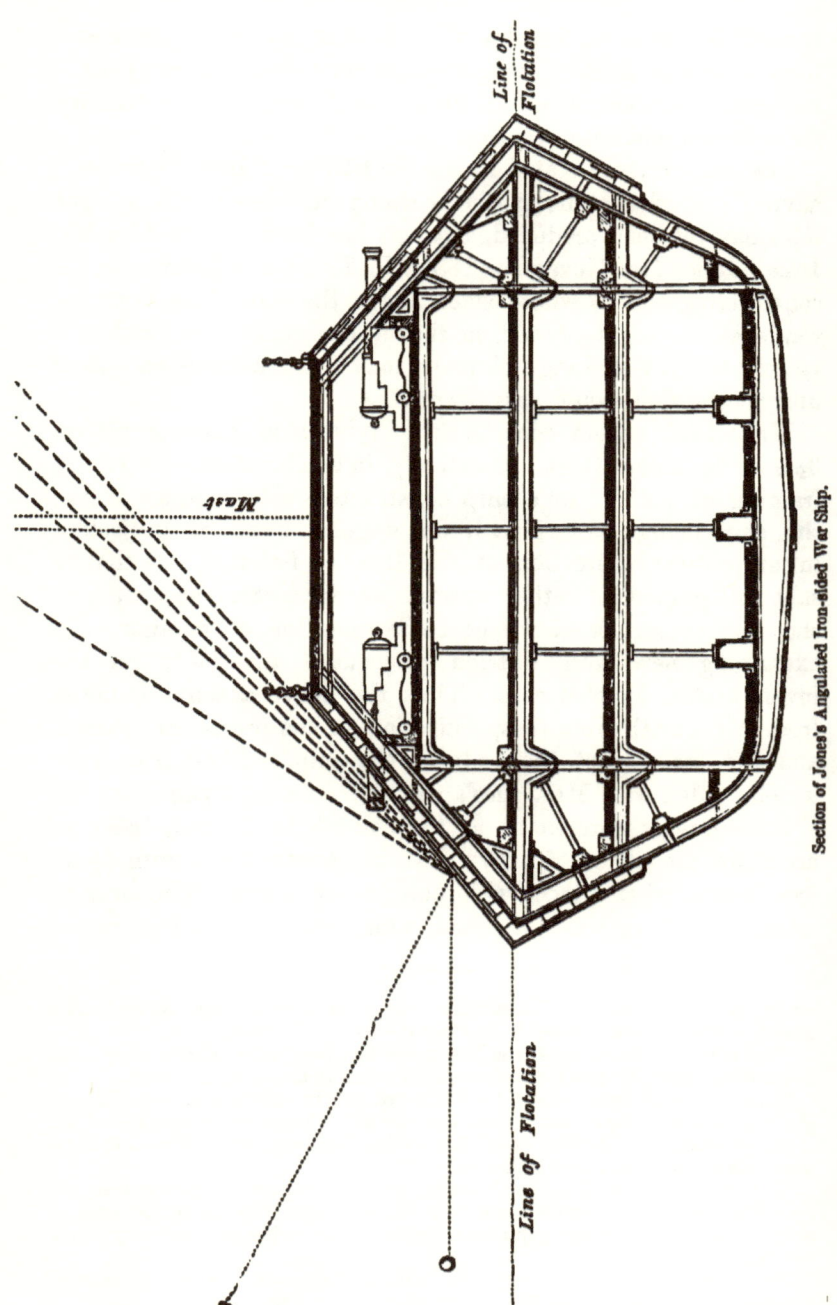

Section of Jones's Angulated Iron-sided War Ship.

sent a portion of the side of a floating battery, was constructed in the following manner :—

The timbers forming the interior of the battery were made of ½-inch iron plates, 21 inches in depth, and placed 14 inches apart at the outside, as well as at the top and bottom. The plates composing the timbers were connected together with ½-inch iron plates to their edges (which were turned at right angles for this purpose) by means of nuts and screws. Outside of the iron plate connecting the timbers was laid stout fir planking, 13½ inches deep, upon which the armour-plates were secured.

The plates were secured by means of wrought-iron bolts passing through the fir planking, and secured with screw-nuts bearing against the interior of the ½-inch plate connecting the timbers. The whole thickness of the side, measuring from the surface of the plate to the inner edge of the timbers, was 39 inches. The plates, four in number, each measuring 7 feet by 3, were placed side by side, with their narrow ends against the upper and lower edges of the butt; thus covering a space measuring 12 feet in length by 7 in the slope, the vertical height being 4½ feet.

No. 1 plate was composed of 4½-inch steel, supplied by the Mersey Ironworks.

No. 2, 3½-inch steel, supplied by the same Company.

No. 3, 4-inch wrought iron, supplied by the same Company.

No. 4, 4½-inch wrought iron, supplied by the Derbyshire Company.

The butt was placed upon a very strong and solid foundation, formed by heavy balks of timber, and supported in its angular position by four stout oak stanchions secured to the balk forming the foundations by iron knees; also further supported by six 3-inch iron stanchions, the heels of which butted against a shoulder formed by a heavy balk of timber firmly fixed across the whole length of the foundation. The heads of these, as well as the oak stanchions, rested against a ½-inch iron plate, which connected the upper part of the interior edges of the timbers, the slope of the stanchions forming an angle of about 100° with the butt. The lower part of the butt was further supported by a ridge of timber firmly bolted to the foundation.

The butt on its foundation was placed on the upper deck of the " Sirius," the bulwarks of which had been cut down to the waterways to enable the lower part of the plates to be seen from the gunboat.

The plate manufactured by the Mersey Ironworks lasted 17 blows in a space of 5¼ feet by 2½ feet before any piece of it was removed, and then the iron was not effectually penetrated, nor the woodwork behind it much injured.

The steel plates, as usual, proved very brittle; and after a few shots it was considered of no use continuing experiments upon them.

The bolts securing the plates to the woodwork were very close together and not uniform, measuring from 12 to 18 inches apart; the bolt-holes proving, as has always been the case, the weakest part of the plate—the cracks generally extending from one hole to the other.

The deck, from the sides falling in so much, must necessarily be much reduced and cramped, so that there might be some difficulty in getting the muzzles of the guns beyond the port-sills. From the falling in of the sides to the extent of 45° or 52°, as in the Portsmouth experiments, the breadth of the gun-deck upon an angulated iron-sided vessel is so much reduced as to be incapable of receiving and working her guns, and it is necessary to give such vessels a far greater length of beam at the water-line, so as to obtain a gun-deck of the ordinary breadth. Here at the outset is a very formidable objection. The proposed upper works cannot be applied to the bottoms of the very numerous existing ships, and therefore new bottoms of the necessary width at the water-line must be provided at an enormous cost, without considering how such an increase of beam must act against the theory of corresponding length and beam, and all the other qualities of a sea-going vessel. Some idea of the enormous cost of ships on Jones's principle may be formed from the fact that the cost of the side angulated plates would be from 36l. to 40l. per ton, to be added to the expense of forming the body of the ship.

The firing took place from a 68-pounder gun of 95 cwt., with a charge of 16 lbs. and cast-iron shot, at 200 yards from the butt. The result was that the shot on striking broke into numerous fragments: these were deflected up the inclined plane over the ship and fell into the sea to the distance of 200 yards, the spread of the fragments extending over a considerable surface of the water. The same results were obtained throughout the experiments. The angulated side was not penetrated nor the timber-work by which the plates were backed-up injured; for the horizontal force of the shot being resolved into the component forces, one perpendicular and the other parallel to the oblique plane, shows how much the penetrating power of the shot in a direction perpendicular to the plane was reduced. The great force with which the fragments of the shot or the shot itself were deflected up the inclined plane in a cone of splinters as shown in the annexed diagram (p. 20), proves that such a vessel could have no masts; for masts, spars, sails, and rigging would be inevitably destroyed: or, if a percussion-shell should be used, the deflection of the explosion would set her rigging in a blaze, and therefore such a ship could not be employed for sea-going services. In the experiments here alluded to the fragments of the shot deflected upwards, passed over the ship where the rigging would have been had she been rigged, and fell several hundred yards beyond her on the friendly side into the water.

It was proposed to try the effect of wrought-iron shot; but this could not be done in harbour, for the shot deflected up the inclined side would have described a pretty considerable secondary flight. It is clear from this that the effect of covering land-batteries with angulated plates of iron at the same inclination would deflect the shot or its fragments into the interior of the battery or into the country which it was the purpose of the coast-battery to protect or of a floating-battery to cover. A battery, whether a ship or a land-battery, can scarcely be called shot-proof unless it stop the shot. But here the shot deflected upwards would destroy the ship's own masts, rigging, and sails, if she had any, and do great execution on the side which the ship was to cover. That vessels of such a form are totally unfit for any open-sea service is plain. In anything of a sea, bodies of waves would rush up the inclined plane in a state of surf—as we see waves rush up the long slope of a beach or conducted up the long slope of a break-water, sweep over its crest, and carry everything off the pier. Just so would it be with respect to the angular iron-cased vessels. These serious and dangerous defects in this expedient are fully admitted, and it is to remedy them that Captain Coles, of the Royal Navy, proposes the plan which I am about to describe, and of which a diagram is given in the following page.

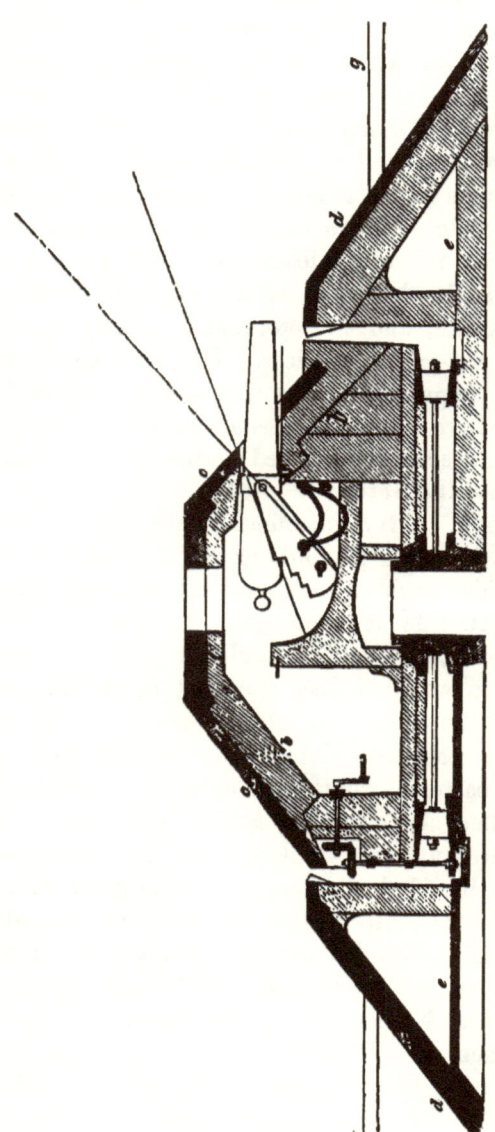

Section of Captain Coles's Circular Tower revolving on turn-table, proposed for the upper decks of angulated iron-sided ships of war.
Diameter, 19 ft. 4 in.

The insertion of lines issuing from the port and passing through the centre of the trunnion is to show the angular space up which the trail of the carriage is thrown by the recoil, bringing the breech of the gun into violent contact with the top of the tower.

With respect to what has been suggested, that, as Mr. Jones's angular-sided ships cannot be employed as sea-going vessels—having no masts and no stability—they might be useful as floating-batteries for coast defences: this brings them into the category, which I have considered elsewhere, of the inexpediency of applying floating-batteries of perishable material instead of land-batteries formed of imperishable materials.

A vessel of this form must be very deficient in stability and bearing. Her bearing is at the water-line diminishing instead of increasing in proportion as she rolls, which is just the reverse of the case from which stability is derived; as, when the bearing or extreme breadth is above the water-line, the vessel displaces with her bilge more weight of water by the side which rolls in than she takes out of the water by the bilge which rolls out.

Here too it appears that the effect of shot fired at such vessels horizontally, from batteries *à fleur d'eau* only, is provided for; but a battery placed high upon a cliff, or other commanding point sufficiently elevated, firing directly shot which either strike vertically or so obliquely as not to glide off, would be just as if the battery and the ship were in the same horizontal plane; and it must be remembered that, even if the shot from a well-placed coast-battery did not penetrate, the force of the blow in the direction of the line of fire would be felt downwards in the body of the ship and throughout the structure—which accounts for the severe injury sustained by the body of the "Sirius," though the target erected upon and firmly bolted to her, and sustained by strong iron knees and stout timber stanchions, was not penetrated. It is a great mistake to suppose that because the side is not penetrated, that the enormous weight of blow is lost, or totally expended upon the side. It is felt throughout the structure, be it a ship or be it a work of masonry. The driving back the target represented in fig. 43, p. 397, 'Naval Gunnery,' 5th edit., the total weight of which was 30 tons, is proof that the force of the blow upon the target, though the shot did not penetrate, was not absorbed by the resistance it met with. The target was erected, but not fastened, on sleepers or beams of great strength; and at every blow it was driven back some inches, and at the end of the experiment was found to have receded several feet.

These very serious defects Captain Coles proposes to remedy by withdrawing the armament from the gun-deck, where in truth there is not sufficient room for it, and to instal two guns in round towers (see diagram, page 24), formed of strong timbers covered with 4½-inch iron plates; and places seven or nine of these towers on the upper very narrow deck—each tower erected on a base which is made to turn upon its centre like the turn-table at a railway station.[10]

The diameters of the railway turn-tables are:—the smaller, 12 feet 3 inches; the larger, 13 feet 3 inches. These turn-tables are well balanced, revolving upon a central pivot, having iron rollers running on iron racers below and near the circumference. The length of a truck, or of a passenger-carriage consisting of three compartments, is greater than the diameter of the turn-table; but this gives the advantage of greater leverage in turning the table. There are larger turn-tables for turning engines. The weight of a truck, loaded or unloaded, being nearly equal throughout, its centre of gravity is very nearly or exactly over the centre of motion; and, however loaded, may be easily turned. The diameter of Captain Coles's revolving towers is 19 feet 4 inches, and the turn-table on which they are placed must be of that diameter likewise. The weight of armour-plating in each tower is about 35 tons, that of the timber framing and platform is about 25 tons: but the weight of the two guns and their carriages, amounting to at least 10 tons, being placed close to the sides of the tower, the centre of gravity of the whole structure is not over the centre of motion; and therefore these towers are truly excentric, and to turn them would require greater power, even were they placed on a turn-table on land.

But how will it be in a ship at sea and in action, even with the smallest floating motions, but particularly when there is any swell? It would require a force far greater than a winch worked by hand, as shown in the figure given in Captain Coles's pamphlet, to turn them, and to suddenly stop the motion of so great a

---

[10] This application to circular batteries is not new: it was adopted in the Maximilian towers at Verona, as may be seen in the article on the intrenched camp at Verona, in my ' Fortification'; but the experimental trials against a tower proved the defects and dangers of such a revolving platform, " The gun-carriages on the tower were destroyed, the platform was crushed, and the whole battery of the tower rendered unserviceable " (' Fortification,' pp. 144, 145).

weight, as we see by the concussion felt in a railway turn-table when the motion is stopped. But there can be no stops at given intervals in laying a gun by the motion of the whole tower: the power applied to turn the tower must always be in gear to move either way by some description of rack-work. When the inclination or roll of the ship is towards the side where the guns are placed, it would be difficult to prevent the centre of gravity from descending to the lowest point; and when the roll or inclination is the reverse, it would require very great power to lift the guns or to turn the tower the other way. The total weight of one of these revolving towers, including the guns, is 68 tons;[11] and Captain Coles proposes that there should be nine of these towers in one of his ships, the total weight of which would be 612 tons' top-weight, in addition to the weight of iron in the angulated side, and to that of the upright iron sides which he proposes, consisting of thin iron plates of sufficient thickness to resist the stroke of a sea and prevent the water from rushing up the angulated side. By this expedient there is a further considerable addition to top-weight of iron covering; and, further, although the armament is the same, yet being placed higher as in the towers, the enormous top-weight of guns towers, and plates, exclusive of other objections, must, I think, prove fatal to the scheme: such a vessel could scarcely swim.

To the weight and cost of iron, in the revolving towers, must

---

[11] The approximate weight of the revolving shot-proof tower, suggested in Captain Cole's pamphlet, is as follows:—

	Tons.
Armour plating	35
Wooden timber of platform	25
Two guns, each 4 tons	8
Making	68

tons for the weight of each circular tower with its pair of guns. The cost of 315 tons of iron plates for covering the nine towers at 37*l.* per ton is 11,655*l.*

When I heard that Captain Coles had published his correspondence with me respecting the estimate which I had given of the weight and cost of iron in his towers, I thought it would be fair to all parties—namely, the public, himself, and me—to obtain the consent of the firm by which that estimate had been confidentially made, to publish their names as the authority on which I had inserted the note in the 1st edition of this Postscript. To this they objected decidedly, as likely, under the circumstances, to expose them to controversy with Captain Coles. They, however, revised their calculation, on the elements given by Captain Coles himself, in the letterpress and figures contained in his work; found the original computation correct, and, moreover, that an estimate from another party made the weight of iron in the towers about 50 per cent. greater than that given in the above estimate of approximate weight.

be added that of the nine iron cylinders which form the pivots
on which the towers turn, each formed of wrought iron $4\frac{1}{2}$ inches
thick: the interior diameter being three feet, and the length at
least $10\frac{1}{2}$ feet; and the weight of each of the nine cylinders
being 18 tons 18 cwt., and the cost 290*l*. But, irrespective of
weight and cost, my objection is taken to the principle of
placing the armament of any ship, whether a sea-going vessel or
a floating battery, in nine separate towers erected on turn-tables
on her decks, each tower containing two heavy guns; to lay
which a bulky and ponderous body, erected on a turn-table,
must be moved as if it were the traversing-platform of an
ordinary gun. This is the principle which Captain Coles proposed, and which I condemned; and no fair trial of his scheme,
for the defence of the great ports by fifteen such ships, can
fairly be tested by experiments made with one small tower,
and that but a one-gun battery.

The failure of Captain Coles's plan must therefore bring with
it the abandonment of Jones's angulated iron-sided vessels, the
defects of which Captain Coles's plan was intended to remedy.
Whatever merit, however, there may be in the principle of
angulated sides is due in priority to Captain Adderley Sleigh,
by his patent of July, 1858, whereas Mr. Jones's patent is dated
1st Nov. 1859.

But it has been said that, although Jones's angulated vessels
cannot have any masts, and, on account of their instability, cannot be sea-going vessels, they may be very useful as floating
batteries for coast-defences. To this there are two very formidable objections:—First, the difficulty already shown, of working
the revolving towers in which the guns are placed, and the great
unsteadiness of such vessels, even in comparatively still water;
and being only batteries *à fleur d'eau*, they cannot therefore
command the sea and defend the coast as efficiently as batteries
placed on elevated sites. Secondly, that of the mixture of naval
and military resources, as a general principle, for coast-defences:
a land-battery, once constructed, may in peace be dismantled,
the guns dismounted, the carriages placed under cover—if of
perishable materials—and the whole be left in charge of a gunner
or two. But floating batteries must be constructed by ship-
wrights; they must be kept afloat and ready, and must be

provided with efficient crews, furnished with naval officers and seamen of any ships that may happen to be in port, or from the Naval Coast Volunteers in the vicinity, which, by so much, would take from the naval service that of which we are most in want, namely, men for the navy.

The 'Report on National Defences in the United States' observes, that "Forts can be made impregnable against any naval force that could be brought against them, and are needed for the protection of our fleets while preparing for hostilities on the ocean. The Government and people of the United States view not with favour the substitution of floating batteries for permanent land-defences, on account of the perishable nature of the former, and the inefficient state in which they may be when sudden danger menaces. The value which they might have, if in perfect order at the moment of being wanted, ceases as soon as the occasion which called them forth no longer exists ; and their speedy decay is certain. To leave the defence of harbours and other permanent establishments to temporary constructions so costly as ships, which are formed of perishable materials, would be to expend enormous sums in a manner which would invite attack by sea. If we rely for defence on our naval force, no portion of it should be permitted to leave our coasts for the protection of our foreign commerce, in the event of an alarm of war occurring.

" To employ our active navy, in whole or in part, for defence, instead of strengthening our fortifications and raising new ones, would be to supplant impregnable bulwarks by perishable ones— a fixed security by a changeable one; it would be to expose ourselves to the chances of being suddenly left for a time without adequate defence. In so doing, we should resign our sense of security and our confidence of safety ; we should divert our navy from its highest duty, deprive it of its chief honour and its chief claim to the respect and support of the people ; we should lose the power of vindicating the national honour and independence, and of asserting the freedom of the seas. The navy is not a defensive, but a protective force."

Our sagacious friends, in the above 'Report on National Defences,' by all the resources of the defensive art, do not say one word about iron-plated vessels. Their conclusion is that,

" Among the various propositions made for the defence of naval
arsenals and maritime places in general, floating-batteries made
of iron so thick as to be shot-proof have been recommended;
and, in order to test the value of such constructions, a target,
representing the side of an iron floating-battery, was formed
with seven thicknesses of boiler iron, well bolted and rivetted together. A shot from a heavy gun passed, without difficulty,
through the target, and tore out large fragments." [12]

The Reviewer states, in page 565, "It is clear that in many
places iron only will be used for sea defences against ships.
General Totten, of the United States' army, guessed that this
would be the case some years ago, and his prophecy seems
on the eve of fulfilment." Now it is clear, from what I have
stated in the Section on Iron Defences in the 'Naval Gunnery,'
and from a mere inspection of the diagrams taken from photographs at the time, that the slabs of iron, if that material only is
used, must be upwards of 8 inches thick, and that the prophecy
of General Totten is not on the eve of fulfilment; for his proposition to combine iron with masonry, to protect the throats of casemate embrasures in masonry defences had been tried, was not
successful, and never will be fulfilled; for the very worst combination that can be made of materials for defensive purposes
is that of stone and iron, which, from their rigid, brittle qualities,
act vehemently on each other, and shake the whole fabric so
formed. *Not verified by experiment.*

Captain Halsted states, in the letter which appeared in the
'Times' of the 22nd September, that "of the three shots fired
from Mr. Whitworth's 80-pounder rifled gun at Shoeburyness,
at the 'Trusty,' one only entered the ship." [13] This is incor-
rect. The first shot struck the edge of two plates and, passing
through them, buried itself in the wood, the butt-end of the
projectile being 11 inches from the surface : the projectile
having struck the end of a bolt the point was deflected upwards. The fore-end of the plate started out from the side

---

[12] 'Report on the national defences of the United States,' 1852, page 6.

[13] The gun used in the experiments was a 10 ft. 4 in. gun, weighing 80 cwt.,
with a 9 ft. 4 in. bore, having 1 turn in 100 inches. The projectile was of
homogeneous iron, weighing 80 lbs.; and the charge 12 lbs. with the first
projectile, and 14 lbs. with the rest.

to the extent of 1¼ inches; about 2 feet from the blow both plates started out and the inside planking was a good deal damaged. The diameter of the hole was a little greater than that of the projectile, and there were no cracks near the point of impact. *This shot did not, therefore, enter the ship*

The second shot struck the plates 16 inches' distance from the former blow, and 4 inches inside the edge of the plate: passing into the ship's side it carried away and broke an iron knee, and fell about two-thirds of the distance across the deck, carrying many wooden splinters with it. A triangular hole was cut out of the plate, of 10 inches by 4. *This shot entered the*

The third shot struck 4 inches from the upper edge of the plate and, passing through it, stuck in the ship's side; the inner end pointing upwards and the outer end projecting about 2 inches outside the outer surface of the plates. It was found that this projectile had struck the butt-end of a beam by which its further progress was stopped. The inside framing of the ship was slightly shaken. The plate was broken in an irregular fracture 8 inches by 8. *This shot did not enter the sh,*

The fourth shot struck about 2 inches from the lower edge of the plate and passed through the ship's side: having gone through a timber, it carried in pieces of the plate and splinters of wood and threw one of the bolts to the opposite side. The projectile made a clean hole in the plate, 6 inches in diameter. *This shot entered the s.*

Perhaps a more formidable statement against the penetrating powers of the shot is that in which Captain Halsted asserts that there is a great discrepancy between the experiments made at Shoeburyness and those which took place simultaneously at Portsmouth, in which the sides of the "Alfred," and at other times the "Undaunted" and the "Sirius," were reported to have been penetrated, as stated in articles 389, 452, &c., of the 'Naval Gunnery,' 5th edit. Captain Halsted asserts that old age had of itself already brought those ships to the verge of the breaking-up dock.

The character and sufficiency of these tests having thus been impugned, there is no doubt that Captain Hewlett would have vindicated the accuracy and sufficiency of the tests which he made and reported to the Admiralty, had he not been restrained by Admiralty regulations prohibiting officers on full pay from writing in newspapers. Captain Halsted appears to say in his letter that the Report of what took place at Portsmouth came to a wrong conclusion. He leads readers to infer that the experiments were made against plates hung to the sides of ships ready to tumble to pieces; but, knowing something of this matter, I venture to say that neither of these allegations is correct. Both the "Alfred" and the "Sirius" were very far from being in such a state. No ship in Portsmouth Ordinary could have been sounder in planking and timbers than the "Alfred." The plates were firmly bolted to the sides. Instead of the results of which Captain Halsted speaks as being totally different from each other in the experiments at Portsmouth and Shoeburyness, the contrary is the fact; but Captain Halsted does not seem to understand that at Portsmouth the practice is invariably made by firing at the same spot until the plate and ship's side have been breached. This has not been done at Shoeburyness, and thus the apparent discrepancy as to the results of breaching and not breaching the sides of the ship is explained. But I think Captain Halsted will admit that the method practised at Portsmouth is the right one, and that it should be adopted at Shoeburyness likewise; for firing single shot at different points is not the way to make the tests, up to the point of breaching.

In the letter signed "A Captain, R.N.," which recently appeared in the 'Times,' the writer appears to think that he has decisively settled the question in favour of iron-sided ships, by stating that the severe damage sustained by the British fleet in the bombardment of Sevastopol on the 17th of October 1854, would have been prevented had the sides of our line-of-battle ships been protected by iron plates. But "the Captain, R.N.," who says he served there, seems not rightly to understand the case. The fire which did such damage to the British ships, and from which they were forced to withdraw, was not horizontal fire, but plunging direct fire from the Wasp and Telegraph batteries placed on the summit of the cliff; and against which those ships could not have been protected unless their *decks* had been covered with shot-proof iron plates. According to this very erroneous conclusion, published on the authority of an officer—present as he states in that affair—we see how it has been run away with and carried to the credit of iron-sided ships. This being so, it has been asserted that "La Gloire," for example, might be laid alongside of one of our coast batteries, and sweep all the artillerymen from its platform, without sustaining any material injury in return. But against batteries placed as all coast batteries should be (see Section on Coast Defences, p. 353, 'Naval Gunnery,' 5th edit.), this would be the reverse. "La Gloire," or any other such vessel, would have her decks penetrated, ripped up, men swept off by shot and shell, that would penetrate into the body of the ship, whilst the battery, in the commanding position in which it should be placed, could receive very little injury in return.[14]

In another letter from a "Naval Officer," which appeared in the 'Times' of the 26th September 1860, it is stated, that, "knowing what he did of the ship attack on Sevastopol in 1854, he felt convinced that the French Emperor had hit the right nail on the head, in producing ships whose sides were coated with iron as the only means of attacking land-batteries." But that expedient would not protect ships against batteries placed in commanding positions such as those that compelled the allied fleet to withdraw. The writer of that letter observes, "with respect to shot passing through the gun ports of a floating battery, mechanical means might easily be invented to prevent this occurring for the future in close action"! In that case the port must be as thick and as strong as the side—the gun could not be reloaded if a mouth-loading gun. A breach-loading gun (Armstrong's) could not be used, because the gun descends to its position from the recoil. This expedient is in either case impracticable. If it were effectual to prevent shot coming into a port, would it not be equally so in preventing shot from being fired out? It certainly is of great importance to keep out the enemy's shell, as observed in the emphatic exclamation of a naval officer, "For God's sake keep out the shells!" This the Reviewer asserts in pp. 558, 559, is "singularly easy," it being "proved that iron plates of $\frac{5}{8}$ of an inch, or, at all events, 1 inch thick, will stop any shell." But the writer is mistaken: it appears

---

[14] See opinion of M. Richild Grivel, quoted on p. 38 *infra*.

by the table on page 84 of this Postscript, that nothing less
than plates 2 or 2½ inches thick will keep shell or their fragments out ; and this would do little for the protection of a
ship unless the plates be thick enough to keep out solid shot
likewise. Covering our ships with plates of that thickness would
greatly reduce their speed, which I affirm is greater than that
of "La Gloire," and so deprive them of the advantage of closing
for action with the foe.

The observations made in the 'Quarterly Review,' p. 558, that
British seamen would shrink from the danger to which they
would be exposed from percussion-shells in a ship unprotected
by iron plates, and that the seamen "have a right to demand of
the nation protection from destruction which seems inevitable,"
is defamatory of the spirit of British seamen, and calculated
.o create rather than to prevent panic. British seamen would
dread no danger from shells, and particularly from the rare
occurrence of a time-fuse shell taking full effect upon a ship in
horizontal firing, provided she had speed to carry them into
close action, instead of being screened under the shelter of the
iron sides of a slow vessel ; and it is defamatory of British
seamen to say that they would prefer the shelter of the iron
sides of a slow vessel, which would compel them to an action of
"long bowls" and distant firing, and deprive them of the
power of being brought into close and terminal struggle with
the foe, which was their wont of old, and will ever be the
characteristic of British seamen.

I have stated in Article 272, p. 267, 'Naval Gunnery,' 5th
edition, "That shell-guns and shell-firing are as yet untried in
actual combat ; and that it remains to be seen what the result
will be on real service." [15]

Reviewing the effects produced by shells on the ships of the
allied fleet, in the bombardment of the 17th October, 1854, at
Sevastopol, it does not appear that the effect of shells upon ships
was so fatal or destructive as had been imagined previously to
that affair ; and I must confess that I partook of the belief that
a shell—and particularly a time-fuze shell—lodged, and then

---

[15] For effect of shells against iron plates see Table of Experiments at
Shoeburyness, at the end of this Postscript, p. 84.

exploding, in a ship, could scarcely fail to set her on fire or destroy her. But from the account given in Art. 366, and Appendix C, 'Naval Gunnery,' 5th edit., it appears that all the ships of the allied fleet were struck and penetrated by numerous shells, yet no ship was destroyed—none "converted into lucifer-matches." Four or five time-fuze shells burst on board the "Albion," and set her on fire several times. The "Sans-pariel," "London," "Albion," and several other ships were penetrated by shells, and suffered considerably in their hulls from shot and shells fired into them from the Telegraph and Wasp batteries. The "London" was three times on fire —the "Queen" forced to withdraw, having been set on fire by red-hot shot. The "Agamemnon" suffered severely from the enemy's shells. One shell burst on the main-deck of the "Arethusa," and carried off nearly the whole of two guns' crews; another shell committed great injury on the lower deck. The "Ville de Paris" received a shell which blew away part of her poop-deck, killing and wounding a great number of men. She received 41 shot and shells in her hull and nearly as many in her masts and rigging; but was not put hors-de-combat, though she was set on fire by shells several times, the fire being promptly extinguished.

Admiral Bruet Willaumez states in his work, 'Battailles de Terre et de Mer,' that the "Ville de Paris" was in a state to repair all these damages, and ready to go into action again.

The affair of Sinope was much more serious ('Naval Gunnery,' p. 312, 5th edit.): the whole of the Turkish squadron was burnt by firing time-fuze shells into them, by which they were set on fire from the ignition of powder circulating in the fighting-decks, and which there is no doubt produced so much panic among the crew that they were unable to extinguish it. There was no panic on board the British and French ships, most of which were several times on fire but easily extinguished.

These well-established facts do undoubtedly very much reduce the estimates which have previously been formed as to the fatal effects of shell-firing, and show that a timber ship would not be infallibly converted into lucifer-matches by shells,

c 2

and certainly not by percussion-shells. The writer of the article in the 'Quarterly' enumerates correctly the different denominations of shells and other incendiary projectiles, but he does not seem to understand their specific faculties and modes of acting. He states that the most destructive shell to ships is a percussion-shell, which explodes either in the side or between decks, tearing up everything in its neighbourhood, throwing splinters about, knocking over guns and men as if a mine had exploded on the spot. He thus confounds the time-fuze shell and the percussion-shell. The effects of a mine can only be produced by a time-fuze shell, which, having lodged in a ship, then explodes—the maximum effect that a shell can produce on a ship. A percussion-shell produces greater effects upon the crew than upon the ship, and is not, like the time-fuze shell, efficient for incendiary purposes : it passes through the side of a ship, in the act of bursting, so quickly, that it has not time to ignite the timber through which it is passing, but is driven in a cone of fragments across the deck, sweeping off the men. The French do not prefer, nor use in naval warfare, percussion-shells—to which the writer attributes the fatal faculty of converting ships into lucifer-matches : they prefer the effects of a mine, which time-fused shells only can produce. Our ships are supplied with percussion-shells (Moorsom's). In battles with iron-cased ships the great object must be to endeavour to send shells through the enemy's gun-ports. And here mark the difference : a percussion-shell thus entering a ship without striking, loses all the effect of passing across the deck in a cone of splinters, and on striking the opposite side with great velocity bursts in passing through the timber, and throws no splinters back. The time-fused shell passing through an enemy's port, lodges in the timber of the opposite side, and there exploding, drives all its fragments back over the deck : the line of least resistance of the mine being inwards. Percussion-shells are moreover very uncertain in their action, for the reasons stated in Article 301, pp. 300, 301, ' Naval Gunnery,' fifth edition ; and the defect does not appear until it is proved by the bursting of the shell at the muzzle of the gun, which shows that the shell had become explosive by a single shock—some concussion having produced the first step by which those shells are rendered explosive. In the recent experiments against the Martello tower at Bexhill, seven of Moorsom's percussion-shells burst in succession near the muzzle of the gun: this being so, it is most important that the whole of that method of obtaining explosion on the second shock impressed on the shell, viz. that of striking, should be overhauled.

The alleged fact asserted in pp. 559, 560, of the 'Quarterly Review,' relating to the attack of Kinburn, as a proof of the complete success of the French floating batteries on that occasion, is not true. Examining the circumstances under which the operations were carried on, it will be found that this fact is by no means established. The ramparts of the bastions constituting the fortress were but little elevated above the level of the sea, and were armed only with from sixty to seventy guns, which were 32-pounders of 75 cwt. each, and these were mounted *en barbette*. The attacking vessels were three in number ; they were covered with wrought-iron plates 3 inches thick, and each was armed with sixteen 50-pounder French naval guns. The ships were stationed at distances varying from 700 to 800 yards, and the firing was kept up during several hours before the place surrendered.

Now the Russian guns were of comparatively small calibre, and therefore they could not be expected to produce great effects on iron-covered ships at such distances ; yet their shot is said to have deeply indented the plates ; and there being no merlons to protect them, it is not surprising that more than half their number were dismounted, and that nearly all the rest were disabled, so that they could not have been again fired. If there were no casualties on board the floating batteries, this can only be taken as a proof of bad gunnery on the part of the Russian artillerymen, who appear to have been very bad gunners in not having so skilfully aimed their guns as to throw their shot through the ports. Had their batteries been placed on elevated sites, as were those of the Telegraph and Wasp batteries at Sevastopol, the French floating batteries would have inevitably been torn to pieces by plunging direct fire.

My account of the bombardment of the 17th October, 1854, contained in Appendix C, p. 616, was drawn up on information sent to me by the gallant Admiral, Commander-in-Chief, and by some of the Captains of ships-of-the-line engaged in that operation. The draught of that account was sent to several of the latter for revision before it was published. Its exactitude was fully admitted, and so emphatically by a gallant officer whose ship suffered much from the Telegraph and Wasp batteries, that he stated that if I had been present at the operation I could not have described it more correctly.

The very aspect of the Wasp and Telegraph batteries, on elevated sites suitable for the purpose of plunging fire, as laid down by me in the Article on Coast-batteries, and for which reason they were denominated in the flag-ship and others the "Douglas batteries," showed pretty plainly what might be expected from them in any encounter with ships.

This being declared so on the part of the British Admiral and several captains of ships, let us see how what I have written stands as to exactitude in the opinions of French naval officers. In a work published by M. Richild Grivel of the French navy, in which my account of the bombardment of Sevastopol is adopted by the writer, we find the following confirmation on the part of our gallant allies, in addition to that of our own naval officers, of the perfect accuracy of my description. "The forts and floating batteries of Sevastopol fulfilled entirely the conditions which the learned General of Artillery, Sir Howard Douglas, proposes for the coast defences :—difficulty of approach for vessels, caused by banks and other submarine obstacles; earthen batteries (*en barbette*) on elevated sites; casemated stone forts for works on the level of the sea; guns of formidable calibre and long range; howitzers and mortars of great calibre; a concentration of preponderating cross and *plunging fire* on all the surrounding zone of the sea." [16]

---

[16] The same writer makes the following remarks on the kind of fleet necessary to dominate the ocean :—"A la flotte de guerre, vaisseaux, frégates et autres bâtiments d'un grand tirant d'eau, appartiendra toujours la souveraineté sur l'océan et dans les eaux profondes. Une nation qui renoncerait à ces premiers représentants de sa force militaire, pour ne plus construire quelles bâtiments de flotille ou de transport, destinés à des usages purement spéciaux, serait infalliment rayée de l'échelle des puissances navales."—' Attaques et Bombardements Maritimes,' par M. Richild Grivel, (p. 69).

The floating batteries, and those which we were urged by the Emperor Napoleon to construct for service in the Baltic, were utter failures. They could not carry their own armament, and all their guns were conveyed in transports hired for the purpose ('Naval Gunnery,' fourth edition, article 'Floating Batteries'). One of those we constructed was sent to Bermuda, as a floating battery for the defence of the coast; and I have no doubt is giving proof of the objection so well stated by Mr. Sidney Herbert in the House of Commons, of the perishable material upon which her armour is placed. Should her services be required a few years hence, if she exist at all, experience will show her ineffi- ciency as a floating battery from having no " command " over the sea. The sites of all our coast-batteries, and those of our insular possessions and colonies, should be revised, and elevated sites preferred as the most effectual against any ships, and par- ticularly those whose sides are protected by iron plates, whilst their decks are seen into, and may be fired into. This revision of coast-defences, say at Bermuda, is required for this reason,— that batteries have hitherto been placed to oppose the advance of sailing-ships; but steam-ships might find out many weak points accessible to them, though not so to sailing-ships, and which have therefore not been fortified. I have reason to believe that this will be found to be the case to a very considerable extent at home, and more particularly in our insular possessions and colonies.

Creeks or intricate approaches which it has been hitherto, for the reasons above stated, dangerous or impossible for sailing- ships to attack, being now accessible and assailable by steam- ships, will become points of attack. This subject requires imme- diate consideration, as will be found when the matter is fully inquired into.

With respect to the *frégate blindée*, " La Gloire," the corre- spondent of a contemporary, quoted by a writer in the ' Army and Navy Gazette,' tells us that the plates with which she is covered have been exposed to the severest test for five years, and that these plates are not iron. He says,—

" In the first place, these plates are not iron, but an amalgam of iron, steel, and another substance, the nature or name of which I have, notwithstanding the most strenuous efforts, been unable to ascertain; but the composition is much lighter than iron, enables the thickness of the plates to be immensely increased, while their impenetrability to shot, conical or otherwise, has been fully proved. The experiments took place in this fashion :—A target was formed with great care, exactly like the section of a ship's side, and was covered with these new plates. It was then for six months fired at three times a week at various ranges, the maximum of which was 100, and the minimum 25 yards, by unrifled 90-pounders throwing a round shot, and by rifled 50-pounders throwing a conical shot weighing 100 lbs. One shot only (a conical one) penetrated the plates; the head of the ball lodged in the plate, and the concussion was such that the remainder of the shot broke clean off as if it had been shaved by machinery."

If the above statement be true, it is clear that "La Gloire" would be covered with the thicker, more resisting, but lighter plates stated to have been discovered and proved five years ago; but it is certain that "La Gloire" is covered with wrought-iron plates $4\frac{1}{2}$ inches (English) thick. With respect to the plates having been proved by firing at them unrifled 90-pounder solid-shot guns, it is only necessary to say that no such gun exists in the French service. The French naval guns consist of 30-pounders of different dimensions, numbered 1, 2, and 3. The 50-pounder was introduced into the French navy in July, 1849, but withdrawn from the naval service, after sufficient trial, on account of its great weight and length; but it has been restored to the naval service for the armament of the floating batteries which were used at Kinburn, and for the armament of the *frégates blindées*. Of these "La Gloire" carries 36. The weight of the 50-pounder is 91 cwt. English; the length, 121·34 inches; the charge, $17\frac{3}{4}$ lbs.; the diameter of the bore, 7·64 inches; the diameter of the shot, 7·44 inches. The canons-obusiers, which enter largely into the armament of the French navy, generally known by the term Paixhans guns, are of three classes, of which it is only necessary here to describe the first, which was 9 feet 4 inches long, weighing 71 cwt, and intended to fire hollow shot or shells, but incapable of firing solid shot efficiently on account of the small charges which the chamber could contain —the diameter of the bore of the gun being 22 centimètres

(8·65 inches English), whilst the diameter of the cylindrical chamber is only equal to that of the bore of a 24-pounder gun French. These canons-obusiers being what we call shell-guns are incapable of firing solid shot, and there being no solid-shot 90-pounders in the French service, it follows that the whole of the statement above referred to must be fabricated. I am enabled, from my own judgment and knowledge, to assert that the whole of the above extract is fiction, pure fiction, concocted by a person utterly unacquainted with the subject—either made on hearsay or unwarrantable assertion, or coming from some interested projector. An amalgam must essentially contain mercury, for it is a combination of mercury with other metals; but the mercury gives brittleness and weakness to the compound containing it, and it is much heavier than iron. A combination of metals, to answer the purpose of iron, cannot be made so much lighter than iron as to allow its thickness to be immensely increased; and that plates have been exposed to the severest tests of mechanical resistance to shot for five years, and that such plates are to be called new plates, is incomprehensible. I therefore pronounce that the whole description contained in the paragraph I have quoted, is made up of exaggerated, false, or interested statements, imposed as truths upon the credulity of editors and readers.[17]

As to the question of rendering iron more tenacious and capable of resisting the penetration of shot without the interpolation of timber, we must rest satisfied for the present with what we know of it experimentally, and not reckon on what we may come to do with it by properties not yet discovered. When the discovery is made the application will follow; we can now only avail ourselves of it in its present state. We are, I think, just as likely to find some substance better than iron as to discover in iron new properties not yet known. It is, no, doubt, well to look forward to the possibility of combinations or new compounds; but it will not do to wait for a discovery which may never be made.

---

[17] Lest I might not be considered sufficient authority on the extra-professional subject here treated of, I think it right to say that the above observations were submitted by me to the highest authority in the world on such subjects, Professor Faraday, and completely confirmed by him.

It is not easy to get at the truth respecting " La Gloire" : no French naval officer will be anxious to discover and to disclose any imperfections in the Emperor's pet plan, reproduced, as has been stated, from the time of Napoleon the First, and no French publication will venture to impugn it ; but I know the truth to be, that " La Gloire," if not an utter failure, is not a success.

The body of " La Gloire" is said to have been modelled on the lines of the " Napoleon," of 91 guns and of equal displace ment ; but, if so, a much greater weight is put upon it than any vessel of that class formerly carried. The armament of the " Napoleon," as regulated by decree of 1849 was—

Lower deck	{ 4 canons-obusiers of	74 cwt.	=	296 cwt.
	28 30-prs. No. 1, of	59 „	=	1682 „
Main deck	{ 6 canons-obusiers No. 2, of 60 „		=	360 „
	28 30-prs. No. 2, of	49 „	=	1372 „
Quarter deck and forecastle	24 30-prs. No. 3, of	30 „	=	728 „
		Amounting to		4438 cwt.

The armament of " La Gloire" consists of thirty-six 50-pounders of 91 cwt., amounting to 3276 cwt. This, together with 820 tons of iron plates, is a weight far greater than the armament of even a first-rate ship-of-the-line of three decks ; but " La Gloire," being a corvette carrying 36 guns, is of much greater length than the " Napoleon," and hence the great amount of armour which she requires.

I assert then that " La Gloire" is a failure ; that she is so over-loaded with armour and armament, that in anything like heavy seas she not only takes water into her ports, but that the sea rolls up her sides and over her ; that she pitches very heavily in a head swell from want of buoyancy to ride over it, as might be expected from being heavily loaded with armour at the bow and stern, where the weight is not supported by displacement directly under it, but mainly by longitudinal strength ; that her speed has never realized anything like that which was expected, for that, instead of being 13½ knots, it has never much exceeded 11, although in her experimental trips she has not had upon her all the weight that would be required for service, excepting coals, of which she can only stow sufficient for seven days' steaming ; that she could not fight her main-deck guns in a sea in which our first-class frigates would be comparatively at rest ; and that therefore " La Gloire" is a very bad gunnery ship, her rolling motion being great and quick, so as in a great degree to vitiate the precision of her rifled guns. When launched and fitted for sea, it was found that she did not carry her guns quite six feet above the water,[18] and she was very deficient in stability. I require not to be told this—it is demonstrable.

No steam ship is fit for ocean service unless she is provided with full sailing power; and no fore-and-aft rigged vessel is capable of acting under sail in a fleet, because it is only by the processes explained in Article 90, page 93, and Article 94, page 96, 'Naval Warfare with Steam,' of backing and filling, by bracing-by or shivering the top-sails, and top-gallant-sails, if set, that vessels in a fleet under sail can be kept in line and position, or perform any evolutions under sail. The sailing power of "La Gloire" consists of her three fore-and-aft lower sails, three gaf-top-sails, jib, fore-stay-sail, and main-topmost-stay-sail, and, when going free, a square fore-sail, and her utmost speed under sail is very limited. A square-rigged vessel, with her courses furled in action, is not likely to have her sails set on fire on the discharge of her own guns, but a fore-and-aft rigged vessel, if under sail, must have her lower sails set, and so be exposed to the great danger of having them set on fire, and, when in action under steam, must have the whole of her deck crowded with the inflammable lumber of sails which cannot be furled aloft.

Those who are conversant with the principles upon which the equilibrium of a floating body depends, know that in a position of equilibrium the pressure of the body downwards, that is, its weight supposed to be applied at the centre of gravity of the ship is equal to the pressure of the fluid upwards; and that the nearer the meta-centre approaches to the centre of gravity of the fluid displaced, the greater the instability of the ship from defect of pressure of the fluid upwards to restore the equilibrium which any alteration in the position of the vessel had disturbed. In proportion as the meta-centre approaches to the centre of gravity, the equilibrium—which is stable when the meta-centre is above the centre of gravity—becomes unstable or indifferent when the meta-centre and the centre of gravity coincide, as

---

[18] See subsequent confirmation of this, note [34], p. 82.

when the floating body is a cylinder; and when the meta-centre is below the centre of gravity the ship will upset.[19]

On the above principles it is clear that in " La Gloire," bur-thened with the weight of her armour of 820 tons put upon her (computed from the area covered, the plates being $4\frac{1}{2}$ inches thick), the meta-centre must be so near the centre of gravity as to deprive her of much of the stability which a good sea-going ship must possess.

" La Gloire " was no doubt calculated to be an efficient ship in speed, stability and capacity, as a sea-going ship, when not loaded as she now is with a prodigious weight of armour ; but by which she is converted into a comparatively slow, unstable, and inefficient vessel for ocean service. This is not a change in the right direction—this is not a triumph of skill over brute force.

But it must be observed that, in vessels of greater magnitude, displacement is increased in a far greater ratio than the external surface which is to be covered with iron plates ; and, if the ves-sel of increased magnitude be similar in form to the smaller, the displacement would increase in the triplicate, and the sur-face to be covered with iron in the duplicate ratio. The capacity of a ship will thus be increased for receiving an engine of increased power, and for stowage of coals. Thus the Admiralty did and do right in not confining the ships they have laid down to mere imitations of " La Gloire," in which those quali-ties do not exist, but by constructing larger ships in which these requirements may be obtained. But it must be observed, that as the consumption of fuel " varies with the vessel's draught of water and with the cube of its velocity—a double velocity being produced, *cæteris paribus,* by an eight-fold quantity of fuel "[20]— so great a space is requisite for the stowage of fuel that in the case of such vessels as the " Warrior," so loaded with armour and armament, the capacity available for fuel is insufficient for a sea-going ship.

Pursuing the consideration of vessels of the displacement of " La Gloire," so much has been said of the comparative strength of iron-cased and uncased ships, and of the utter

---

[19] 'Naval Gunnery,' note Art. 459, 5th edit.
[20] 'Naval Warfare with Steam,' Art. 67, p. 59.

inability of the latter to contend with the former in combat, that perhaps it may not be without use to make a few general remarks on the circumstances of the case, with respect to vessels of nearly equal displacement: the one covered, like "La Gloire," with iron plate; the other, like one of our first-class frigates, unprotected by iron.

Speed and metallic protection are antagonistic properties, and cannot be combined in a ship of that displacement. Two vessels equal·in every respect to each other in speed will not have equal speed when one is loaded with armour in the manner of "La Gloire," and the other not so loaded, as in the case of the "Mersey." The ship so overloaded must lose speed, and consequently be deprived of the power of choosing the distance at which it is most advantageous for her to fight; but the other by being unloaded with armour will retain the speed with which she was endowed, and so be able to choose her distance from the foe. Metallic protection and power of speed are therefore antagonistic qualities. One or other of these must be sacrificed. Which of them should be given up, is a question which I propose to examine hereafter. In the mean time it may be remarked that it does not appear, on a careful review of improvements and changes in the art of war, that the best way of opposing new modes or means in the practice of war is to imitate those innovations. If this were so, the practice of war would not have altered, as we see in a review of its history. It would rather appear that improved science seeks to counteract rather than imitate; and the question arises whether this might not be successfully done in the case of iron-sided vessels, which must necessarily lose speed. It is therefore a question whether, by superior speed and a judicious mixture of armament—68-pounder guns on the main deck, and Armstrong's long-range 40-pounders on the upper decks—a vessel such as our "Mersey" might not be capable of contending with such a ship as "La Gloire," with her speed reduced to inferiority by the heavy armour under which she labours.[21]

---

[21] " The importance of superior speed cannot be too highly estimated : of two ships of war which may be equal in every other respect, that which has the greatest speed has a decided advantage over the other, and in an action is most sure to win. With respect to the relative efficiency of steam and wind,

I do not believe that it is the intention of the Emperor of the French to proceed in forming a fleet of *frégates blindées*, on the type of "La Gloire," for ocean service, as the writer in the 'Quarterly Review' asserts His Imperial Majesty had determined to have ready in the spring of 1861, in order to urge us to stop forthwith the building any more timber ships, and immediately to set about constructing an iron fleet to protect the country against that of France, preparing and ready to appear at the time specified. We have arrived at that period, and a solitary vessel, "La Gloire," has been produced—and she a failure. It appears, therefore, that we need not be in any haste to commence this revolution in the navy of England by converting it into a navy of iron ships.

Superior steam power will give to steam fleets that advantage which superior sailing-speed gave to ships in times gone by. In the case now under consideration, superior speed, even a fraction of a knot per hour, is of vital importance, and should be sought for by every possible contrivance. This, as I have stated elsewhere ('Naval Warfare with Steam,' pages 74 to 77), has been obtained by my improved propeller, the performance of which shews that the change proposed by me is in the right direction, and it was pronounced by the inspector of steam experiments to increase speed, diminish vibration, and improve the steerage of the ship; yet no trial has been made of my improved propeller, as I most earnestly requested might be done in my letters recently addressed to the Comptroller of the Navy.

It is a sage maxim in war, or in preparing for war, not to overrate your own force nor to underrate that of the actual or expected enemy. What, then, can be the reason which induced the Emperor Napoleon III., his Minister of Marine, and the 'Moniteur de la Flotte,' to reverse that sage and very trite maxim by vaunting the superiority of "La Gloire," and thereby underrate the force of our ships. The reason is this: it is a ruse, a decoy, to frighten us into discontinuing building any more timber ships of the line, and to apply our pecuniary

---

it is impossible any longer to regard the unsteady and uncertain power of the wind as more than an auxiliary to be occasionally employed in subordination to steam, and chiefly for the sake of economising fuel."—'Naval Warfare with Steam,' p. 102.

and naval resources forthwith to construct our channel and ocean fleets of iron-cased corvettes like " La Gloire ;" and this ruse has had the desired effect of inducing a portion of our press (as we see in the pages of the ' Quarterly Review') to be duped and so delude the country.

A very able and distinguished French officer, M. Richild Grivel, a great admirer and advocate of the Emperor Louis Napoleon's floating-batteries, constructed for the attack of forts and fortresses and other special purposes in inland seas, condemns, in no measured terms, the notion of the practicability of using them in ocean fleets as substitutes for line-of-battle ships. Admitting fully the advantage of floating-batteries, gunboats, and other vessels of small draught of water, for the special services above stated, he has well said that " to dominant fleets of line-of-battle ships—the true representatives of naval power—for service in the open sea will always belong the sovereignty of the ocean ; and that the nation that would renounce these true representatives of naval power, by constituting their fleets of comparatively small ships, adapted only to services purely special, would be infallibly erased from the category of first-rate naval powers." [22]

If we should be so infatuated as to commence forthwith the reconstruction of our navy on such a principle, England would assuredly, ere long, be erased from the category of first-rate maritime powers, and lose the empire of the seas. Such really has been the delusion in the case of " La Gloire," such the panic upon the mere appearance on the sea of that solitary frigate,—and she a failure,—that I should not be surprised when that delusion shall have been dispelled,—and it is passing away,—if it were spoken of as in the days of the Warner hoax. I am quite sure that as much nonsense, deceit, deception, and credulity are exhibited in the one as there were in the other. But, though I know that " La Gloire " is a failure in speed and in all the qualities required in a sea-going ship, yet the public mind has been brought to such a state of fever and delusion on this subject, that I will allow that the Government is perhaps right in laying down other such monsters before the " Warrior," the " Black Prince," and the " Defiance," are launched, and have been

---

[22] See note [16], p. 38 supra.

tried. Yet I really thought it would be prudent to launch and try them first, for, if successful, that would at once put us far ahead of the French iron-cased ships, of which only one exists and she a failure, and that, should the aforesaid ships now building be failures, that we should have the advantage of experience in correcting their defects in our future constructions. With respect to the "Warrior," "Black Prince," and "Defiance," I am happy to find that the scheme of fitting them to act as steam-rams is abandoned for the reasons stated in the 'Naval Gunnery,' pp. 430-3, and in the 'Naval Warfare with Steam.'

I have hitherto noticed in this Postscript the subject of wooden vessels of various forms covered with iron plates, and now proceed to the subject of vessels constructed wholly of iron, as insisted on in page 562 and others in the article on iron ships, and the assertion that "the days of timber ships are over." Here the question naturally presents itself, that if this be so, why was not the much-vaunted frigate "La Gloire," which had been held up as a fine specimen of the disuse of timber, been constructed wholly of iron. To explain this the article in the 'Quarterly' asserts (pages 560, 561) that France, being far behind us in the manufacture of iron and in building iron ships, had not sufficient supplies of that metal with which to build iron ships; but that, having ample stores of timber in her dockyards and an army of trained artificers in wood in her arsenals, the Emperor, determining to have a fleet of vessels of that description ready for sea in the spring of 1861, decided on the only course left open to him, namely, to build at once wooden vessels to be plated with iron as a temporary expedient; but that, these being less durable and less safe than vessels constructed completely of iron, that system would be carried out thereafter, according to the discovery made of that great nostrum by Napoleon III. But, as I have already said, the discovery was made by Colonel Paixhans forty years ago, and I deny that it ever was intended, recommended, or deemed practicable in France to form vessels for war purposes entirely of iron, as may be seen by reference to the 'Proceedings of the Institute of France,' confirmed by the experiments made in the United States. The French and United States' naval architects know well that no iron ship could float if formed of iron so thick as to be impenetrable to heavy solid shot. Slabs of wrought-iron 6 and even 8 inches thick are broken through and knocked to pieces by 68-pounder solid shot, as we see by the experiments detailed in the Section on "Iron Defences," in the fifth edition of the 'Naval Gunnery,' and which prove at once that neither for land nor for floating-batteries can iron be applied, unless strongly backed up by timber or by masonry. The photograph here introduced of the

state of utter ruin to which a battery formed of a combination of masonry and iron was reduced in the experiments (Article 407, page 405, 'Naval Gunnery') entirely disposes of that scheme. Iron enters largely into the construction, rigging, and equipment of ships, as in building of houses, bridges, and other works of civil engineering; but the foundation on which the naval architect builds his structure is laid in the sea, by the displacement of his ship, and the problem he has to solve is to displace the greatest quantity of water with the least possible weight of ship; so that the load which she can carry be a maximum. This problem is, to a considerable extent, reversed in the case of iron ships with shot-proof sides, or iron-cased ships—the weight of the material of which they are formed being a maximum.

With respect to forming ships for war purposes entirely of iron, their top sides being 6 inches or, as the United States authorities

have it, 7 inches thick, the Reviewer appears to be utterly ignorant
of the fundamental condition of the doctrine upon which flotation
and stability depend, namely, that the weight of water displaced
by the ship up to the intended load-line must be equal to the
total weight of the floating body and all that she contains.   I
should like much to see the writer's draught of the form and
dimensions and weight of such a ship, with a computation of the
weight of water displaced, the difference between the two being
that weight which the ship can carry at the load water-line.   I

very much suspect that the Reviewer did not understand and could not make a computation of this problem. With respect to the assertion in page 566, that however it may be as to covering or not the sides of ships with iron plates, their bottoms below the water-line should be formed of iron, as being more enduring, less expensive, and better able to resist the shakes occasioned by the screw: this may be safely denied. A well-built timber ship of good oak, copper-fastened and sheathed with copper, is far more enduring, and is kept in repair at less expense than iron-bottomed ships. The bottom of a well-built wooden ship, copper-fastened, seldom wears out; it is a fact well known to all ship-builders that wooden bottoms will last out three tops.[23] The case is the reverse with respect to iron ships; their tops last out three bottoms, which being constantly acted upon by salt water are corroded far sooner than the tops, and soon become leaky by the corrosion of the bolts. The shakes of the screw, or any other concussions, are more destructive on the rivets and bolts of an iron ship than on the caulking of a timber ship, and no copper can be used in conjunction with iron. The bottoms of iron ships soon get fouled, as we see in the case of the " Great Eastern," by which speed is greatly impeded unless such ships be docked and cleansed.

The article in the ' Quarterly ' states, page 566, that so completely had experience proved that all that part of a ship which is below the water-line should be of iron, that no screw vessel of wood belonging to any port in England has been built since the first experiments with the screw were tried! However this may be with respect to private dockyards, which I very much doubt,

---

[23] To prove the lasting nature of timber-built ships, I may state that, by a recent Report to the Admiralty, it is shown that the " Nelson," 46 years old, has lately been converted into a screw line-of-battle ship; that the " Royal George," built 33 years ago, has also recently undergone the same process; that the " Eagle," 56 years old, has until a late period been in commission; and that the following screw-ships are considered effective line-of-battle ships, had, at the date of the Report, the pendant flying, and were about to go to sea on a cruize, viz. :—the " Ajax," built 1809 ; " Edinburgh " and " Hogue," built 1811 ; " Pembroke," built 1812 ; and the " Blenheim " and the " Cornwallis," built 1813. In addition to this, is given a list of sailing frigates now " in ordinary," and of which 16 are capable of being fitted for the conveyance of troops, the ages of which are as follow :—two 52 years old, six averaging above 44 years, eight 34, four above 26, seven nearly 16, and three averaging 12 years ; besides eight of the old 42-gun frigates of 1080 tons each.

this at least is sure, that the whole screw steam navy of England
is composed of wooden ships, and they were built in the ports of
England.   Iron ships for commercial purposes may be constructed, and perhaps with advantage, of thin plates of iron, but
these are totally unfit for any purposes of war, as has been already shewn; and as passage-vessels they are dangerous, as fatal
accidents have proved.   The crew who engage to serve in them
are not obliged to encounter the risk, but troops are compelled to embark in them, as we see in the fatal instance of the
" Birkenhead ;" and the case of the " Royal Charter" is more
appalling.   Passengers entrust themselves in the one case to
incur whatever additional risk there may be, troops are compelled to encounter it whatever be the risk.   No case of failure,
danger or death, has occurred to the Cunard or United States'
steam timber ships, which have been running for many
years.

The 'Quarterly Review' not having taken any notice of my
description, in pp. 404 and 405, of the experiments made
against slabs of wrought iron 6 and 8 inches thick which were
penetrated and broken through by 68-pound solid shot, I reprint
the annexed figure, as an ocular demonstration which, *sautant*

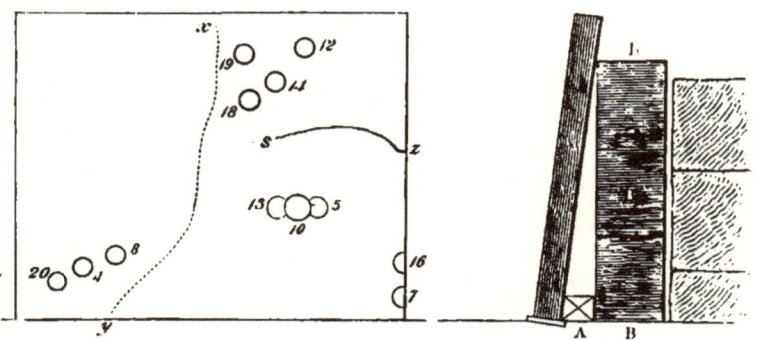

aux yeux, will convince the reader at a glance that the assertion
in the 'Quarterly Review' is absurd, Q. E. D. Yet the Reviewer
persists in stating that the days of wooden ships are over, that
our dockyards for constructing timber ships should forthwith be
converted into fitting ports, and that iron ships should be constructed by contract in private yards, for that (page 565 ' Quar-

terly Review') no officer or ship-builder connected with our dockyards knows anything of iron shipbuilding, and they naturally
feel a distaste for what they do not understand, and cannot help
being aware that when iron ships are introduced the occupation
of such functionaries is gone! The writer of the article in
which these absurdities are chronicled in immortal type will be
expected to explain how he proposes to proceed. Will he construct his iron vessels of plates thicker than those of 6 and
8 inches, which he now sees photographs of in these pages
penetrated and broken through by 68-pound shot? Or, if he
now is forced to admit or see that the weight of such a ship
would at least be very nearly equal to the weight of water displaced, and that therefore iron can only be applied in thin
plates, which, if not backed up by some other resisting body,
would be penetrated by much less powerful shot than 68-
pounders, will he state what is the material to which these plates
could be applied? Can it be to any other material than timber?
If any mistake has been made in the construction of the
"Warrior," it consists in having formed that ship on a thin
iron skin $\frac{5}{8}$-inch thick, instead of on the body of a well-built
timber ship, copper-fastened and copper-sheathed, on account of
the enduring qualities of timber-bottomed ships compared with
the far less enduring properties of submerged iron, as stated in
p. 51 of this Postscript, and which material is specified in the
contract for building the timber iron-plated frigate proposed to
be constructed in England for the service of Russia. And "La
Gloire"—the *beau idéal* of the advocates of iron ships, and held
up by them as the type upon which we should forthwith commence the reconstruction of our navy by discontinuing the
building of timber ships—is a timber ship! Not provisionally,
as the article in the 'Quarterly' states, for want of iron, but
decreed to be built of timber, upon the principle that to do so
with iron is impracticable; so that "La Gloire" is a practical
denial of the system of which the advocates of iron-built ships
hold her up as a model. What then becomes of the assertion
that the days of timber ships are over? The days of timber
ships, whether commercial or warlike, are not over, nor ever will
be over; and especially so with respect to an extensive empire
like Great Britain, possessing colonies and dependencies in every

region of the earth.   And our neighbours on the other side of the Channel have not yet produced even the type of a ship fit for war on the ocean, of which to form the fleet of iron-cased ships which were to convert all our timber-ships opposed to them into lucifer-matches.

The Reviewer having, as he thinks, settled the question as to the efficacy of iron ships, and iron defences for sea-service, adds, pages 564, 565, that iron plates will likewise be used with equal efficacy to protect land-defences against ships—that iron is everywhere carrying the day against its more perishable rival.   This has been sufficiently tested and negatived by the results of experiments in this country.   I have already stated that plates of wrought-iron, 6 and 8 inches thick, are penetrated by 68-pound shot.   But as in land-defences there is not the same necessity for using plates of wrought-iron as afloat, being lighter, trials were made of blocks of cast-iron, 8 feet long, 2 feet wide, and $2\frac{1}{2}$ feet thick, supported in the rear by a rectangular mass, consisting of 6 heavy blocks of granite, each block $1\frac{1}{2} \times 3 \times 2$ ft., leaving $4\frac{1}{2}$ ft. of the centre of the target unsupported as represented in the figures pp. 49, 50, and which were penetrated by 68-pounder solid shot, at 600 yards' distance, and entirely destroyed at the tenth round.   The figure, p. 49, represents the state of ruin to which a target, in which iron was combined with stone, was reduced.

Some interesting experiments were made at Shoeburyness in December, 1860, to test the efficacy of lining the cheeks of casemate embrasures with sheets of iron, and to strengthen the embrasures by masses of iron applied very differently from those in the experiments carried on in the United States, which I have described in Articles 414-16, pp. 409, 410 ' Naval Gunnery.' With respect to the first, the results obtained were, that shot, either whole or in splinters, and grape-shot, were deflected into the battery, and caused great destruction to the wooden men placed in position 'as gunners at their posts: these results conforming entirely to those of the American experiments.

The mode of strengthening the throats, cheeks, and outer edges of casemate embrasures in the United States was by masses of wrought iron 8 inches thick, composed of 16 plates, each ½-inch thick, firmly welded together; the iron forming a sort of frame firmly bonded into the masonry. In those experiments it was found, that, though the impact of shot only indented the masses of iron, yet that the masonry behind was considerably shaken, jarred, and some stones displaced; that the vibration of the wall was very considerable; that the motion of the granite stones was very slight where the target was supported by earth, or where the stones were bonded into a considerable length of brickwork and concrete; but the energy of the vibratory movement of the wall was very great, and might have been measured, had anything so marked been foreseen. In the recent experiments at Shoeburyness, it was proposed to apply bars of wrought iron built up of several thicknesses of iron, morticed into each other, forming a bar 10 feet long, 4 inches deep, and 10 feet long. So far as the iron bar only was concerned, the trials were successful; but, when the bars were in contact with the masonry, it was so much shaken that the Committee proposed to make the embrasures with a square opening of 2 feet square, the bars so placed as to be independent of and separate from the masonry. This proves the accuracy of my remark that a combination of two hard, rigid, and brittle materials—stone and iron—bonded together, is the worst combination of materials that can be made. In consequence of this fact, it is proposed by the Committee to face the whole casemate embrasure with iron bars extending to the whole width of the casemate; no masonry being behind it, and kept in its position by a strong framework and struts of timber, throughout the whole breadth of the casemate. Experiments are about to be tried, to test the efficacy of this mode of metallic protection. It appears to me, I confess—unless the whole of the masonry exposed to fire should be protected by iron—that, although the embrasures might be protected, the piers of the arch might be breached: but this will soon be tested.

The conclusion at which the Reviewer in the 'Quarterly' arrives is, that, however disappointing the necessity, just as we have constructed our screw steam-navy at an enormous outlay, we should be called upon at once to begin to construct another navy of iron ships, *de novo*, from having persevered in the wrong direction, after discovering that wooden ships had been superseded by the only Power whose fleet we have at present any reason to fear. Where shall we find the necessity for this?

When the alarm, occasioned in this country by the threats which appeared in ' L'Enquête Parlementaire,' of invasion of this country was expressed and implied by the sudden and extensive augmentation of the French navy, the proposal to adapt the steam-ships belonging to the packet and commercial navy to the purposes of war was proved to be impracticable, as shown in p. 16 of this Postscript.

Whether our Government may or may not in a future war consent to abandon the belligerent right to detain enemies' ships carrying goods contraband of war, and to relinquish altogether privateering, it is of importance—to an extent that can scarcely be expressed—that the ships of our mercantile marine should be able to defend themselves. But how will it be when the Reviewer in the ' Quarterly ' shall have persuaded the merchants of England to trust their goods to vessels entirely of iron, incapable of defending themselves in war, against the swarm of efficient timber ships, public and private, which other nations will send forth?

I am in a position to explain pretty accurately the views and intentions of all foreign Naval Administrations on the subject of iron-cased ships, and ships made wholly of iron, and shall commence these explanations with a review of the stupendous exertions made in France to regain that position as a first-rate naval power which was first overthrown by the anarchy occasioned by the Revolution, and ultimately revived by the British Navy during the glorious and successful naval war which ensued.

In an article in the ' Révue Contemporaine,' of February, 1859, written by M. Grivel, capitaine de vaisseau, that very distinguished French officer has given an historical sketch of the important foundation laid down by me for the improvement of our naval gunnery, which, though found greatly superior to that of France in the remarkable war with that nation, did not in the sequel appear to maintain its superiority when tried against the vessels of the United States in the war of 1813-1814.

M. Grivel commences that sketch in the following terms :—

"To what hands should a nation, jealous of her maritime preponderance confide the improvement of her naval artillery as well for the defence of her coasts as for the armament of her ships? That is the question which the course of events presented to the consideration of England, when the General of Artillery, Sir Howard Douglas, undertook to resolve it."[24]

From this foundation the whole history of the improvements introduced into the British Navy by the adoption of my system is carefully traced in detail, all my views and arguments fully stated, and the results correctly given as models worthy of adoption in France. My work was translated into French by M. Chappontier, capitaine de vaisseau, and gunnery schools for the training of naval gunners adopted in France, even before my plan was established on board the "Excellent" in England ; and, as I have already observed, more extensively and liberally carried out in France, with respect to the number of trained gunners per gun, than is the case with the British Navy. In a letter recently received from M. Grivel, acknowledging the receipt of the copy of the 5th edition of my 'Naval Gunnery,' that officer states that it is to be found in the hands of every French Naval officer who takes a real interest in the efficiency of his profession, and I have every reason to believe that the new matter contained in that Work relating to iron defences is producing very strong impressions.

It is not easy, as I have already said, to get at the truth respecting the opinion of French officers upon Louis Napoleon's pet scheme of iron-cased ships and fleets for ocean service to be formed of such vessels, for officers are deterred from writing and publishing on such subjects by the necessity of obtaining the official authority required in the French service to make such publications, and by the rule that such papers must be signed by the writer. But I believe that, if French Naval officers were permitted to discuss Louis Napoleon's plan as freely as

---

[22]. "A quelles mains une nation jalouse de sa prépondérance maritime devait-elle confier le service de son artillerie navale, tant sur ses côtes que sur ses vaisseaux? Telle est la question que les évènements venaient de poser à l'Angleterre, quand le général d'artillerie, Sir Howard Douglas, entreprit de la résoudre."—*Revue Contemporaine et Athenæum Français*, vol vii., p. 667.

British Naval officers and others may canvass in this country the measures adopted by the Naval administration, it would appear that few French officers do really approve of the system of iron-cased vessels for active warfare on the sea, and that none approve of forming ocean fleets of such vessels. I need only refer, in confirmation of this, to the opinion of M. Richild Grivel, which I have quoted on p. 47 *supra.*

With respect to my work on 'Naval Tactics with Steam,' a copy of which I sent to Admiral Bruet Willaumez, to reciprocate his courtesy in sending me his work 'Batailles de Terre et de Mer,' M. Grivel states that that distinguished French Admiral who is now Commander-in-Chief at Toulon, has not yet published a new edition of his 'Battailles de Terre et de Mer,' but has expressed to M. Grivel on various occasions his high estimation of my work on 'Naval Warfare with Steam'—a treatise spoken of in terms of high commendation in French periodicals and reviews, and which would no doubt be noticed by the Admiral in a new edition of his work. And I have reason to believe that the French fleet which is about to proceed to practise steam tactics, and to familiarize officers and seamen with the evolutions of steam fleets in the open sea, will be exercised in the formations and evolutions contained in or taken from my work on that subject; whilst it remains, I think, a dead letter, or nearly so, in our Naval Service.

With respect to Russia, my pamphlet put a stop to the iron-cased frigate about to be constructed in this country, a telegram from St. Petersburg having been dispatched to that effect just as the contract was on the point of being signed. I am well known in Russia as the author of the 'Naval Gunnery,' which was translated, by command of the Emperor, into Russ on its first appearance. But I am better known as the son of the British naval officer who, as an Admiral in the Russian Navy, greatly improved that of Russia, at a time when it was the policy of England to bring her forward as a great Naval power. At the peace of 1763 my father, who then commanded the "Tweed" frigate, was paid off; and, being already known as a scientific and excellent officer, received a proposition from the Russian ambassador, with the authority of the British Government and

Naval Administration, to undertake for a time the rank of rear-admiral in the Russian service, in order to promote the improvement of their naval system in all its branches. This position he accepted; and, being authorised to select an officer to assist him in his mission, he took with him his former first-lieutenant of the "Tweed," Greig, with the rank of captain. My father remained in Russia for four or five years, when, being appointed at his own request to the command of a line-of-battle ship—I think the "Stirling Castle,"—he resigned his situation as Admiral in the Russian service. Captain Greig succeeded him; and from that distinguished·and accomplished officer, whose letters to my father I have often perused with deep interest, have sprung those distinguished officers, the Greig's, whose names are so conspicious in the Naval Annals of Russia. This sort of hereditary claim, together with the translation of my ' Naval Gunnery' into Russ, opened to me those sources of information on which I ventured to warn the country in 1854 that all that had been said in disparagement of the Russian naval system, and particularly of their gunnery, was erroneous, as we felt afterwards at Sevastopol.

The ' Naval Gunnery' was translated into Dutch by a very able officer, Captain Gobius, who afterwards obtained the distinguished position of Minister of the Navy, and died about three years ago. Through the acquaintance, however, which I formed with him, and an extensive correspondence with him and with others, I am enabled to say that the opinion of the great bulk of the Dutch Naval officers is that iron ships are unfit for purposes of war; that some small steamers, destined for India, have been constructed to trade with countries where facilities for cleaning and repairing may be found; but no iron ships are provided for other purposes: that perhaps they may construct some iron-cased ships, in case of necessity, for coast defences, though not fit for general service as sea-going vessels; but in general that iron applied to the construction or protection of vessels is in no favour in Holland. So far as my communication with Denmark, Sweden, and Norway[25] enable me to judge,

---

[25] With regard to Sweden and Norway, I know that the naval authorities believe it to be better to allow experiments, for the solution of the question of

there is no indication in either of those naval services of their adopting either iron-cased vessels or iron vessels for purposes of war, though there is no lack of iron there.

The only exception—and it is a remarkable one—to the foregoing statement of the views of all foreign naval powers, other than France, on iron-sided ships, is that of Spain. Two types of "La Gloire" are constructing by the builder of that ship, at Toulon, for the service of Spain. A large screw-steamer, of wood, iron-cased, is contracted for to be built in this country, from which also tenders are invited for the construction of two iron-cased, timber-built gun-boats for the Spanish Government; and ten more iron-cased gun-vessels, similar to those of France,· are building in Spanish dockyards; possibly to act in combination with the French iron-cased vessels, on some future occasion. Thus Spain is again becoming a respectable, if not a formidable, naval and even colonial power. The disruption of the American Union is benefiting Spain, as we see by the restoration to her of the sovereignty of the island of St. Domingo. Spaniards have been emigrating in considerable numbers from Cuba to St. Domingo, with a view to its re-annexation to Spain. There is probably no alliance more courted by France than that of Spain, in Louis Napoleon's Eastern policy and Mediterranean ambition. And we see that the connection which the Emperor has formed with Spain, by marriage, will be cemented by political alliance with that country. Let us not forget the purpose for which floating-batteries were invented, and squadrons of gun-boats provided, for a great operation which lasted from July, 1779, to the peace of 1783. In securing our military stations abroad, Gibraltar should be put in a state of defence, seaward, without loss of time, and armed abundantly with the new long-range guns, and with all the missiles and incendiary weapons best adapted to resist floating batteries, in positions commanding what must ever be the weak point of iron-sided ships.

---

iron ships and armour plates, to be carried on at the expense of other nations, while they confine their own attention to securing *a powerful artillery distributed among small vessels,* considering the configuration of their own coasts, which alone they have to look to the security of, and not having, like Great Britain, to provide for the defence of colonies and to maintain a naval prestige.

It has already been observed, page 18 of this Postscript, that in the Report of the Secretary of the Navy of the United States not one word is said either of iron ships or iron-cased ships. This ignoring of the question so hotly disputed in this country, does not arise, as some have asserted, from the plenty and cheapness of wood and the lack of iron, but from disapproving in principle of that mode of construction.

Our transatlantic brethren have peculiar views and objects of their own on all matters of improvement; and the subject of iron-sided ships and iron ships, which is now in fierce discussion in England, they consider to involve costly experiments which they may well leave us to work out, whilst they confine themselves to the manufacture of machinery for the propulsion of their steamers, and to building wooden ships. These being articles of home manufacture and growth, timber is getting remote and scarce in the great ship-building state ; but, as the iron-masters of the Union could not compete with those of England, the· supply of iron for casing or building ships would have to be obtained from England, if the United States Government were in favour of that principle of construction—a thing not to be tolerated where protection is so stringent by duties which are collectively called revenue duties, but which are nevertheless protective. And this is the reason for the lamentations of our iron-masters, that the Americans are not wise enough to build iron-cased ships for purposes of war nor packet and mercantile ships of iron. The Naval authorities of the United States will certainly not adopt iron-cased ships[26] for ocean service, or for coast defences. If iron-plated ships should ever be adopted by them, it would only be in the event of war on the inland seas, the lakes of America ; but, even in such a case, they would confine themselves to plates of sufficient thickness only to keep out shells. They would, as seamen have always been wont, brave round shot, and, with superior speed and good tactics, fight the battle out as of old.

---

[26] " The ' coat-of-mail ships,' now *bugbearing* the world, will prove wholly impracticable as cruisers ; although, for special service against a neighbouring belligerent power, they may no doubt prove effective : more particularly if ever it turns out that they are made impervious to heavy shot."—' Steam for the Million,' *Ward*, page 99.

Assuming that then foes could be equally protected, a writer judiciously observes that no decisive results would be obtained by either, as in the operations on Lake Ontario in 1813 and 1814, between Sir James Yeo and Commander Chancey, when the naval superiority on the lake was not obtained by fighting but by building ships against each other; and therefore was only a question of money, just like that which is draining France and England of immense sums of money in building iron-cased ships against each other. Some naval, and other officers there are in the United States, of high authority, who think—what may be considered as rather a quaint view of the subject of protective armour to ships—that, instead of keeping shells out, they should let them in through bulwarks so slender that percussion-shells would not receive a sufficient shock in passing through them to produce explosion, and would therefore pass the deck like shot and find their way out; and that time-fuse shells, unless they burst at the very instant of impact, would likewise act only as shot. There is something worth considering in these two postulates: the first would dispense with the construction of huge vessels, such as the " Warrior," with sides proof against any shot; and the second would let all projectiles pass through, and so defeat to a very considerable extent the action of shells. How such vessels are to be armed and managed we are not told, but it seems to have engaged some attention.

In the United States, then, we see no indication of the adoption of iron-cased ships, all their packets and all their merchant ships are of timber. They think that iron bottoms cannot be preserved, even with frequent docking and painting, for more than five or six years. They know well the catastrophes and the dangers which we have so fatally experienced by iron-bottomed ships. If, for the purpose of economising wood and iron, the United States should adopt the system of compound ships—*i.e.* ships compounded of wood and iron—they would make their bottoms of strong solid timber capable of resisting and preventing the penetration of rocks, and form their top sides and bulwarks of thin iron. This is the reverse of our system: we form the bottoms of our merchant-ships of iron, their top sides and bulwarks of timber; thus placing these materials just where they are most liable to perish.

One of my correspondents, speaking of "La Gloire" and the "Warrior," states that, "as cruisers they are the veriest absurdities conceivable;" that "the endeavour to give a great mass like the 'Warrior,' high speed by sharp lines and great steam-power is akin to turning a siege-train of 24-pounders into a flying battery by hitching on an increased number of horses, or to expect that a race-horse carrying additional weight could go at the same rate as when not so loaded."

Condemning thus our large compound ships, the "Warrior," &c., I am asked by one of my correspondents, a distinguished officer of the United States navy,—Why might not that part of such ships which is made shot-proof, forming a quadrilateral portion, be deprived of its unprotected ends and applied as a floating redoubt or refuge in the Mediterranean or the Gulf of Mexico, where there is, for the most part, good anchorage ; and so form a refuge, where a certain number of light and active vessels might find protection and thus maintain naval superiority in that sea ?

I have stated in the 'Naval Gunnery,' pp. 424, 427, 433, my opinion on the subject of steam-rams, to which I would refer the reader. The opinion entertained, I think, very generally by naval officers of the United States may be inferred from the following article in Commander Ward's work 'Steam for the Million:'—" The steam-ram must be of enormous weight, strength, and cost, and may be of some service about home, but will very likely turn out a *sheepish* affair ; she might perhaps sink the vessel struck, but she would take the ram down with her by the horns. In war, defence keeps pace with the attack ; and the adoption of steam-rams would introduce appliances for grappling with them before the victim can sink, so that when the ram endeavours to extricate herself from the stricken ship by reversing the engine she will find her horns caught in a thicket. It is easier to get into a scrape than to get out of it. Will the rams carry their weighty bows in a heavy sea, or will iron-cased ships carry their heavy armour as cruisers ?" I have said in the 'Naval Gunnery,' that attempts to storm vessels by boarding will be frequent in steam warfare, and should be provided for accordingly by requisite facilities for giving and receiving the assault. To board a ship should not be attempted by stemming her like a ram, but by running up alongside, the men swarming on board the enemy's vessel along her whole side when lashed to the foe. To board a vessel, end on, would be like the error of storming a work in a deep column under a narrow front instead of attacking it in line throughout its practicable points supported by reserves.

With respect to the 'Naval Tactics with Steam,' I may be permitted to refer to the subjoined note[27] from Commander Ward's work on Naval tactics in which my work is incorporated.

Thus we perceive that the system of iron-cased ships is confined, as an exceptional case, to France for aggression and to England for self-defence; and, so long as that mania may be continued, we must be provided with a superiority of such vessels. But when, from whatever cause it may terminate, we shall not be justified in keeping up an expensive establishment of iron-cased vessels for remote purposes of aggression, which England never will attempt. I quite agree with Captain Sulivan that, if Sir Richard Dundas had been provided with half-a-dozen powerful iron-cased ships, he might have destroyed the unprotected floating defences of Cronstadt; but that opportunity has gone by, and certainly will never recur. If we persist in keeping up, unnecessarily, for any eventuality and at enormous expense, squadrons of iron-cased vessels, we may be assured that we shall never find our enemies unprovided with similar defences.

I confess I am not a convert to the system of iron-cased vessels in principle, for general and universal services and purposes; but I am, like many other persons, willing to admit the necessity of adopting that system for our own security, for the *special* and *exceptional* purpose specified, so long as the Emperor Napoleon persists in the aggression which the construction and extension of his iron-cased ships menaces, and

---

[27] " So recent is the introduction of steam into the navies of the world, that no maritime battle, and but little experience of any kind, is afforded from which to draw examples and illustrations. But the language has been graced, and naval science enriched, by Sir Howard Douglas, in his work on ' Naval Warfare with Steam,' to which the reader is referred, as to a mine of professional knowledge. . . . As a landsman, Sir Howard treats subjects which especially appertain to the sea with remarkable clearness and accuracy."—' Manual of Naval Tactics,' by James Ward, Commander, United States Navy, 1859, p. 137.

which is referred to in the subjoined note,[28] taken from page 9 of that able work, the 'Admiralty Organization,' which represents the unwise contest as draining the financial resources of both countries, with no advantage to either; and which he thinks Mr. Cobden, who advised the Emperor to form the treaty respecting free-trade, should have induced him to abandon.

Iron is not imperishable, as stated in the House of Commons, even in air. The roofs of the iron huts at Thorncliffe Camp, though covered at great cost with the well-known lacquer, have perished to a great extent, and are now being replaced by wood covered with asphalted felt.

In the Report of the Commission for constructing the new Palace of Arts it is recommended that it should be built of stone, it having been found that in the Crystal Palace, formed of iron and glass, the former had perished to a considerable extent.

It is well known to timber-ship builders—and surely they should be heard as well as iron-ship builders—that the bottom of a well built copper-fastened timber ship scarcely ever wears out, and at any rate will last out three tops. In iron-ships it is the reverse; one top will last out three bottoms.

Two plates—one taken from the bottom of the "Jamaica," of 421 tons, built at Glasgow in 1854, the other plate taken from the bottom of the ship "G. F. D.," built at Newcastle in 1849—have both been nearly worn through and destroyed from sea-water without and bilge-water within.

I am in possession of some iron plates, taken from the bottom of an iron ship when under repair in a dock, ten years old, the rivets of which had been destroyed by corrosion and broken off flush with the inside lining; and which were actually pushed off with the point of an umbrella, being only retained in their places by adhesion, and when afloat by the pressure of the water, all of which bolts must have fallen out in any strain or concussion which the ship might have received.

---

[28] "Those gentlemen would render a great service to their country, as well as to the cause of economy, if they would induce the Emperor to abandon the intention, which the newspapers attribute to him, of increasing his iron-cased ships from six to nineteen with the greatest despatch; and it might be suggested to them as a means at once of testing their own influence and the pacific intentions of the Emperor."

E

In another case, an iron steamer from the coast of Africa, when taken into dock and the pressure outwards removed, about 300 bolts actually dropped out into the dock. Recently an iron store-ship, being taken into dock at Woolwich, to have her bottom scraped and cleaned, about 500 bolts fell out, the bilge-water having in both cases eaten off the rivets.

The timber ship "Nile" struck on the rocks near Visaga-patam, and lay grinding there for fourteen hours; she came off and proceeded home without making a drop of water, although, as it appeared when in dock, her keel had gone and many of the bottom planks were rubbed nearly through. In 1854 the "Alfred," with a full cargo, outward-bound, met with a similar accident, and when in dock at Calcutta, showed the same amount of damages as the "Nile," but reached home in safety. Her Majesty's frigate, the "Pique," of 40 guns, got on shore on the coast of North America, and remained several hours, but she got off with the loss of her keel, and reached home in safety. Had these vessels been iron ships, all of them would have been lost.

Captain Sulivan, in his speech at the meeting of the Institute of Naval Architects, states that, in addition to the case of the "Pique," on one occasion he came home in a vessel that he commanded, which had never leaked at all, insomuch that they had to pump water into her to keep her sweet; but when she came into dock it was found that sixty feet of her main keel had been ground away, and all the solid bottom, in one place, except a few inches, and yet she was perfectly water-tight to the last.

In reference to the "Great Eastern," I contend she is utterly unfit for any purposes of war: she is neither shot-proof nor shell-proof—nothing less than plates $2\frac{1}{2}$ inches thick will keep out shells and their fragments (see abstract of experiments, page 84); she is not fire-proof, seeing that she set fire to herself on a memorable occasion, and burnt her own saloon, the restoration of which, with other repairs, as we have recently seen, cost the shareholders 18,000*l.*, paid out of profits which are absolutely *nil;* she is not more durable than a timber ship; and should she ever strike a rock or reef, she would crack and break up and be totally lost. When off the Land's End she

But she did not, when on the rocks near M?

rolled to such a degree that it was found necessary to play the engines on the mast-heads to prevent the eyes of the rigging from catching fire by the friction; and should she be armed with heavy guns in the same proportion to her displacement, as the "Ariadne" for instance, she would shew herself an awful roller and utterly unfit for purposes of war.[29]

With regard to liability to fracture, an iron ship is formed of at least twenty times as many pieces as a wooden ship, and the plates of iron are not joined together by bolts so strongly as timbers are by scarfing. This accounts for the fact that when an iron ship strikes a rock, she breaks up into pieces immediately; the bolts broken, the plates themselves torn across like brown paper, as proved by inspection of the fragments of the "Metropolis" brought up by divers, and of which I have given a sketch in the account of the wreck of that ship at Jersey (p. 69).

It has been stated that the "Nemesis," Captain Hall, resisted shot: I happen to know from an officer who was in her, that on the occasion alluded to only one shot struck the "Nemesis," and that it passed through and through her, though certainly not above a 12 or 18 pounder.

The "Queen Victoria" iron steamer of 1152 tons, built at Newcastle in 1856, stranded in Barnpool on the 2nd January

---

[29] In the account of the departure of the "Great Eastern" for America on the 1st of May, the 'Times,' in allusion to the rate of speed on her last voyage having been low to what had been predicted of her, states :—" The deficiency was attributed on all hands to the very foul state of the vessel's bottom. Accordingly when the "Great Eastern" returned, and she became at once plunged in her usual chaos of difficulties—her screw bearings out of order, her decks deficient, and everything wanting more or less of readjustment and repair, it was decided to lay her up for the winter on a gridiron at Milford Haven, and put her into thorough order. The gridiron on which she lay was short of her enormous length by very many feet, and thus a short space at the stern, and a considerable portion of the bows, projected over without support at either end. During some of the high flood-tides of winter, when it was feared she might float, it was found necessary to pump water into her forward compartments. This precaution, of course, brought an undue strain upon the part of the bow left unsupported, and the iron compartment showed signs of bulging and leakage. For this evil there was unfortunately no remedy till the ship was afloat, when the defects were at once made good."

This confirms in a remarkable manner what I have asserted to be her danger, should she, like the "Birkenhead," "Victor Emmanuel," or "Queen Victoria," become at any time suspended by grounding in such a manner that her whole length was not equally supported, viz. that she would break her back, and become a total wreck.

last, laden with the telegraphic cable intended for communication between Malta and Alexandria, and immediately became a
wreck.   Her kelson under the mainmast and the keel were set
up nearly two feet; one joint of her keel 8 ft. 6 in. long, right
under the foremast, was broken ; the stern-post broken and
twisted 10 inches to starboard.   In fact, she broke her back
and became a complete wreck.

I am favoured with the following account of the wreck of the
" Ava," by Captain Haswell of the Royal Navy, who was a
passenger in that vessel :—

"The ' Ava' ran on shore on a rock, off the Island of Ceylon, at 8 p.m.   A
pinnacle rock was forced through the midship compartment, and the engine-
room was almost immediately filled with water, which made its way into the
foremost compartment, and the vessel sunk down by the head until the water
was level with the figure-head (filling the forecastle and forepart of the
vessel), when the fore part rested on the bottom, the after part of the vessel
being raised, and only *very partially* water-borne, the consequence of which
very soon began to show itself: at the upper deck the iron sides of the vessel
began to open, and was rent down to the water's edge as if cut by a saw.   At
daylight, when I left the ship, the opening at the upper deck was a foot wide,
and the water running into the after part of the vessel, increasing the weight
and bringing a greater strain on the after part which was not water-borne.

" The separation of the after part from the foremost was very gradual until
about 4 p.m. (twenty hours after she first took the ground), when the after
part broke short off and sunk in deep water, leaving the foremost part of the
vessel firmly fixed on the rock, with just the upper deck above water.   Only
a moderate breeze and the water smooth, with a slight ground-swell."

The " Victor Emmanuel," an iron-built bark of 520 tons, ran
on shore in Chale Bay at the back of the Isle of Wight, and
went to pieces immediately ; four only of the crew were saved.
She broke up so swiftly that no signal of distress could be made,
as she went to pieces like a glass bottle against a wall.   The
" Victor Emmanuel " was a fine new vessel which cost 10,000*l.*;
the amount of salvage sold by auction was 20*l.*

The following account of the wreck of the " Metropolis " at
Jersey, is extracted from the Public Instrument of Protest made
by Charles Corr, Master.   It states that—

"On the 12th February, 1861, at 4 p.m., when off St. Brelade's Bay, Island
of Jersey, they took a Jersey branch-pilot on board, and steered for the main
roads with the flood-tide and hove to, waiting for water to get into St. Helier's
harbour, the steamer going at quarter speed so as to keep way on her.   At
4. 30 p.m., she struck all of a sudden on the Ronandiere rock, about one mile
and a half to the south-west of Elizabeth Castle; the pilot immediately
ordered the engines to be reversed, but it was impossible to get her off; the
fore-castle, fore-cabin, and fore-hold filling with water, and ultimately the
engine-room ; the steamer at once settling down forward with the bowsprit
under water.   Expecting her to founder every moment, they got their boats

out, sending one of them ashore for a steam-tug to assist them, they remaining near her about one hour and a half in one of the other boats, but finding her settling down more and more, they also pulled for the harbour of St. Helier, to hasten assistance. The screw steamer, the 'Alar,' having her steam up, they prevailed on the master of that vessel to steam at once to the 'Metropolis,' with the intention, if possible, of getting her off, or of saving some part of her cargo. At about 6. 30 p.m., they got close to the 'Metropolis,' which had, however, in the meantime settled down as far as midships; there being a heavy sea, and darkness setting in, the master of the 'Alar' considered it an impossibility to get the 'Metropolis' off, and steamed back to the harbour."

The pilot, George Allix, states that the vessel broke up almost immediately. The diver says that the vessel broke into three pieces; the forepart is still supported in part by the rock on which she struck; the middle portion, containing the engines, broke into several pieces, the engines and shaft fell out; the stern portion is lying where it sunk when it broke off from the fore part of the ship. A beam was brought on shore broken in two; a large piece of a plate, weighing some hundredweights, taken from the bottom of the vessel, is torn up like a piece of paper, the rivets broken, the bolts all out, the iron torn in two in a straight line along the edges of the bolt-holes, as shown in the accompanying sketch.

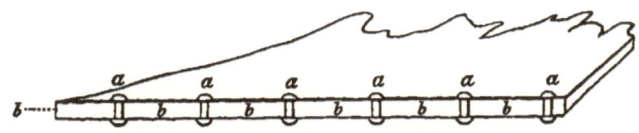

a, a, represent the bolts, the rivets of which are uninjured.
b represents the torn edge of the plate in a perfectly straight line, exposing the bolts.

The troop-ship "Birkenhead," sailed from Queenstown, January 7, 1852, for the Cape, having on board detachments of the 12th Lancers, 2nd, 6th, 12th, 43rd, 45th, and 60th Rifles, 73rd, 74th, and 91st Regiments. She struck upon a pointed pinnacle rock, off Simon's Bay, South Africa, and, of 638 persons, only 184 were saved by the boats; 454 of the crew and soldiers perished by drowning, some of them, perhaps swallowed by sharks that were seen swimming around. The rock broke through into the engine-room, and literally rent the ship in two, the parts sinking on its opposite sides; while those clinging to the wreck calmly resigned themselves to inevitable death.

The following extract from a letter from Rear-Admiral Sir Alexander Milne contains a very authoritative and absolute condemnation of iron ships. "When iron ships strike a rock, as in the case of the "Birkenhead," "Transit," &c., the iron cracks like an egg-shell, the compartments fill forward or aft as the case may be; the portion filled with water sinks; the others float, and the ship breaks up into two or three portions like a well baked biscuit, and such I believe is the case in all such vessels."

When a timber ship is bilged she does not immediately break up, unless, as in the case of the "Avenger," she is overwhelmed by breakers, which no ship could resist; and when a timber ship does break up, the fragments float and many lives are saved.   In a fearful shipwreck which I suffered in October, 1795, in the "Phillis" (the particulars of which are given in the Appendix, p. 89), the ship was bilged as soon as she struck, but she held together for more than 48 hours; and her whole quarterdeck came on shore when she went to pieces, on which the survivors, had they not previously effected their landing, would all have been saved.

The barque "Devonia," Lawson master, from Shields to Aquilas in Spain with a cargo of coke, struck on the south-east spit of the Goodwin Sands, on the night of the 18th December, 1854, during a heavy gale from the N.N.W. very thick with snow. The crew all took to the poop, with the exception of one man, who was washed overboard when the vessel first struck. The vessel broke up, and the poop, with fifteen men, drifted clear of the sand. The "Lord Warden," John Robinson master, on his passage from Boulogne, on the 19th December, 1854, when about twelve miles from Boulogne, at 8 a.m. sighted an object about four miles to the eastward of his track. He immediately bore down upon it, and found it to be the poop of a ship with the mizen-mast standing, with thirteen hands clinging round the mast. There was a very heavy sea running at the time, and it was only after great difficulty that the boat was launched with four men, who succeeded in taking eight men off the poop, on which they had been eleven hours, whilst the steamer "Princess Helena" took off the remaining five; two of the men having died of exhaustion during the night.[30]

---

[30] 'Providential Rescue from the poop of the barque "Devonia," in mid-channel, on the 19th December, 1854.'

When the awful wreck of the "Birkenhead" took place, I sent in manuscript to Captain Chads the observations on that catastrophe and on the dangers of iron ships, which the reader will find in the 'Naval Gunnery,' and in foot-note, page 6, of this Postscript, for the consideration of that high authority; and my much respected friend, in reply to that communication, stated that so fully and unanswerably had I demonstrated the dangers of iron ships, and consequently the decisive objections to employ them as transports for troops, that, if my demonstration of their unfitness for such purposes were published, it would produce such an effect that soldiers, who, he said, were already aware of that danger by the loss of the "Birkenhead," would make manifest their reluctance to embark in iron ships; and he therefore advised me not to publish the article to which this relates, and with this advice I conformed for a time, but the fact which I have related above has so important a bearing upon the question of employing iron ships, either for the transport of troops, emigrants and other passengers, that I consider it indispensable to put in proof the strong sense that is entertained in the army and in the navy of the danger of such vessels. Let another such accident as the loss of the "Birkenhead" or the "Royal Charter" take place, and there would attach to those who, notwithstanding such warnings, have continued to employ iron ships as transports, the most awful and crushing responsibility.

The following account of the loss of the "Assistance" in China, by the master, is given in a letter to Captain Sulivan, the naval adviser of the Board of Trade:—

"I rejoice to find you are fighting such a good battle with regard to the system of employing iron vessels as troop-ships and men-of-war. The 'Assistance' was sufficient to convince me that for all purposes wood must always have an enormous advantage over iron.

"When the accident happened she was making 38 revolutions, which would give a speed of about seven knots.

"The water was perfectly smooth, so she remained completely quiet after striking; and having met with the same misfortune in a wooden vessel, I was not at all alarmed for the safety of the ship. It was low water, the rise being about four or five feet. A stream-anchor was laid out, and we had commenced hauling on the cable, when the carpenter reported five feet of water in the well; and nothing more was done, fearing the ship would go down in deep water if she came off. In four hours the poop disappeared as far as the mizen-mast. No ordinary wooden ship, whether man-of-war or merchant ship, would have been lost through such an accident ; she might possibly have lost a few sheets of copper, or a trifle more damage, and at high water she would have come off. It was the remark of all on board, that they could not have believed it possible that so slight a shock would have been the means of losing the vessel. I cannot but compare the case of my striking on rocks in the 'Vixen' and the 'Assistance.' In the 'Vixen,' we had a gale of wind and a very heavy sea ; we were ten minutes on the rock, and so violent were the shocks that the boilers forced the upper deck up, and the rudder was knocked off, and we expected she would have parted ; but, having thumped over it, she made at first about a foot water in four hours, which we reduced to a foot in twenty-four hours by clearing her out and filling in open spaces from inside. She remained on the station two years afterwards without being docked."

The "Indian" went on a reef of rocks in tolerably calm weather on Nov. 21st, 1859, and parted, the fore part falling over on its beam ends and the after part remaining upright, and went to pieces so suddenly that seventeen men were drowned before the boats could be got out.

Captain Sulivan, spoke as follows in the discussion which recently took place at the Institute of Naval Architects :—

"I thoroughly agree with Sir Howard Douglas in his opinion upon the inexpediency and danger of iron-bottomed ships.

"I think we should not trust a ship like the 'Warrior' upon an iron bottom. Since I have been connected with the Board of Trade, I have examined into the losses of iron ships, and the result of my investigation has been a conviction that iron bottoms are far less safe than wooden ones. I much regret being obliged to come to such a conclusion, because iron is a material which, if it could be safely used, would give this country a great advantage ; but many an iron ship has been wrecked, when wooden ships under similar circumstances would, I believe, have been saved. When an iron ship strikes a rock, the plates seem to give way like brown paper.

"I believe an iron-cased ship, upon a solid wooden hull of large scantling, would have a great advantage over an iron ship like the 'Warrior,' especially when they have to do in-shore service upon an enemy's coast. We are not afraid of knocking away a little wood off the bottom. Such blows as knocked three feet of solid wood off the bottom of the 'Pique,' would have knocked through the bottom of the 'Warrior.' We must not jump to a conclusion with respect to iron bottoms, because iron-ship builders may be honestly of opinion that it is the best material : we ought to have the opinion of the best wooden-ship builders also ; and carefully consider the cases of the ships that have been lost through having iron bottoms. I believe the 'Great Eastern,' if she had struck as the 'Pique' did, would never have come off again.

"The objection to iron ships being, as has been shown, that if they touch a rock, the bottom will be perforated and the vessel lost, applies à fortiori to every iron-cased ship, even in a greater degree, on account of the weight of the ship ; and, if left high and dry, the crushing weight the iron bottom has to support when the outer pressure of the water is taken off ; so that, even if iron bottoms were less perishable than wooden bottoms, which I deny, it is highly inexpedient to form iron-cased vessels such as the 'Warrior,' &c., on iron bottoms.

"From what I have proved, it appears that not only are iron ships utterly unfit for purposes of war as armed ships, but are moreover dangerous as transports for troops and for emigrants. I believe there is not one case in fifty in which the centre compartment will save a ship. I would not like to trust to it myself, and I would sooner have three feet of good solid oak under me than the strongest bottom that iron shipbuilders can put together.

"It has been stated that one reason why iron should be used in constructing ships of the enormous length [31] which we are now producing is the difficulty of getting timber of sufficient scantling to build ships of that great length and tonnage ; but this is carried to a most dangerous extent : the 'Connaught's' length was nine times her breadth, and she, being deficient in longitudinal strength, was lost in consequence. A ship of great length, with her engines placed abaft the centre of gravity, must be liable to great strains in a high head-sea, and require strength of frame at those parts where the heavy strain will be most felt. It has been ably remarked by Mr. Sheddon, on the construction of iron ships, that recent casualties—some of them of a fearful nature—shew that, although iron ships have comparatively speaking proved successful, yet many have turned out to be lamentable failures."

The following iron troop and store ships have been purchased by Government. They have been employed on an average six years, that is, taking the time of four that have been lost up to their wreck.

"Birkenhead" (lost) ;	"Perseverance" (lost) ;
"Megæra ;"	"Transit" (lost) ;
"Simoom ;"	"Adventure ;"
"Himalaya ;"	"Assistance ;" (lost).
"Urgent ;"	

Loss 40 per cent, or 6·7 per cent. yearly.

I have annexed, overleaf, an authentic list of the fearful loss of iron-built ships during the years 1857 to 1861 (May) inclusive.

During the two years of the Russian war we had· 200 wooden ships at sea, most of the smaller vessels, including all the paddle frigates and sloops were on· shore or *rocks* repeatedly in the Baltic, their bottoms being much torn ; but not one failed to do her work after. The only ship lost during the two years was the "Niger ;" and she would have been got off but for the

---

"The mania for increasing length will hardly be cured, until after more dis ster. But unfortunately the victims will be a simple public which knows no better, intent only on going ahead ; and not the capitalist and architect who don't go to sea in the vessels, only order and construct them, under the united impulse of cupidity and vanity."—'Steam for the Million,' *Ward*, page 100.

LIST OF SAILING AND STEAM VESSELS, BUILT OF IRON, BELONGING TO THE
UNITED KINGDOM, LOST IN THE YEARS 1857 TO 1861 INCLUSIVE.
S. denotes Steamers. S. S. Screw Steamers.

NAME OF VESSEL.	Descrip-tion.	Port.	Tons.	Particulars.
1857.				
Madrid	S.	London	315	Off Vigo.
Erin	S.	,,	532	,, Caltara.
Durham	S. S.	,,	428	,, Coast of France.
Bacchante	S. S.	,,	406	Lost.
Niger	S. S.	,,	482	Off Santa Cruz.
Eupatoria	S. S.	,,	435	,, Flamborough Head.
Jessie Macfarlane ..	..	,,	244	In Table Bay.
Abundance	S. S.	,,	328	Off Oesel.
Eagle	S. S.	Hull	423	,, Gothland.
Sydenham	S. S.	,,	314	,, Nargen.
Burlington	S. S.	,,	369	Lost.
Fusilier	..	Liverpool ..	504	,,
St. Andrew	S. S.	,,	824	,,
Sally Gale	..	Newcastle ..	193	In the Scine.
L'Empereur	S. S.	,,	570	Lost.
Briton	S. S.	Southampton	408	,,
The Queen	S.	Aberdeen ..	327	,,
Charlemagne	..	,, ,,	1017	,,
Earl of Carrick	S. S.	Ayr	146	Off Isle of Man.
Clyde	S. S.	Glasgow ..	780	,, Labrador.
City of Madras	..	,,	813	Lost.
Zouave	S. S.	Leith	320	Run down.
Rothsay Castle	..	Port Glasgow	96	Lost.
Ravensbourne	S.	London	402	,,
Admiral Benbow ..	..	Lynn	130	,,
Valdivia	S. S.	Liverpool ..	555	,,
Waverley	..	Glasgow ..	731	,,
1858.				
George Jordan	..	Fleetwood ..	81	Lost.
Madagascar	S. S.	Aberdeen ..	320	Cape of Good Hope.
Vision	..	Glasgow ..	422	Abandoned off Cape Horn.
Thistle	S.	,,	320	On Irish Coast.
Candace	S. S.	London	379	Lost.
Tintern	S. S.	Belfast	58	,,
Catapilco	S. S.	London	43	,,
Ernestine	S.	Grimsby ..	239	,, Certificate recᵈ. April 22.
Cavinsmore	..	Liverpool ..	1085	,,
Rosedale	S.	Newcastle ..	265	,,
Napoleon	S.	Grimsby ..	301	,, Certificate cancᵈ. Aug. 7.
Hunwick	S. S.	Hartlepool ..	333	,,
Don Affonso	S. S.	Glasgow ..	240	Abandoned.
Tempest	S. S.	,,	798	Missing. Cert. cancᵈ. Feb. 19.
Brigand	S. S.	,,	260	Lost.
New York	S. S.	,,	1011	Wrecked.
1859.				
Cape of Good Hope	S. S.	London	420	Wrecked.
Sarnia	S. S.	,,	185	Lost.
General Williams ..	S. S.	,,	955	,,
Paramatta	S.	,,	2166	,,
Vulcan	..	Hartlepool ..	318	,,
Northman	S. S.	Hull	213	,,
Czar	S. S.	,,	740	,,
Jupiter	S.	,,	128	,,
North Sea	S. S.	,,	450	,,
Marion Macintyre ..	..	Liverpool ..	283	Wrecked and condᵈ. Nov. 18.
Anne Baker	..	,,	463	Abandoned.

NAME OF VESSEL.	Description.	Port.	Tons.	Particulars.
1859—*continued*.				
Royal Charter	S. S.	,,	2165	Lost.
Swarthmore	..	,,	923	,,
Preston	S. S.	,,	200	,, off Holyhead.
Prince of Wales ..	S.	,,	279	Sunk in deep water. .
Progress	S. S.	,,	238	Lost.
Emerald	S. S.	Ayr	152	,,
Five Brothers	S. S.	Glasgow ..	398	Off Coast of Africa.
Ignez de Castro ..	S. S.	,,	114	Lost.
Admiral	..	,,	77	,,
Scotia	S.	,,	71	Off Milford.
Carron	S. S.	Grangemouth	239	,,
James Stewart	..	Greenock ..	239	,,
Admella	S. S.	,,	209	,,
Martello	S.	Inverness ..	325	,,
Glowworm	S.	Irvine ..	161	,,
Dunedin	S. S.	Leith	340	In the Elbe.
Junius	S. S.	Peterhead ..	319	Lost.
Empress of India ..	S. S.	,,	451	,,
Mail	S. S.	Dublin	195	,,
Shamrock	S.	,,	163	,,
General Codrington	S. S.	London	371	Stranded.
Vulcan	..	Hartlepool ..	318	Lost.
Flora	S. S.	Waterford ..	456	,,
Indian	S. S.	Glasgow ..	1154	,,
Lapwing	S.	,,	48	Sunk. [sunk.
Volunteer	S. S.	Newcastle ..	222	Struck by the ice at Greenland and
Express	S.	Southampton	152	Lost off Jersey from striking on a
Northam	S. S.	London	904	Near Jeddo. [rock.
1860.				
Scamander	S.	Liverpool ..	667	Abandoned.
Black Prince	S. S.	London	405	Run down.
Newpelton	S. S.	,,	356	{ Sailed from Llanelly, Dec. 29, for Havre, and not since heard of.
Earsdon	S. S.	,,	420	Wrecked.
Malabar	S. S.	,,	623	,,
Admiral Miaulis ..	S. S.	,,	692	,,
Warrior	S. S.	,,	268	,,
Harburg	S. S.	,,	237	Lost.
Connaught	S.	,,	1523	Burnt.
Ganges	S.	,,	217	Foundered.
Vigilant	S. S.	Hull	175	Lost.
Russian	S. S.	,,	598	,,
Midge	S. S.	Liverpool ..	83	Foundered.
Cairn	..	,,	229	,,
Nemesis	S. S.	,,	331	Lost.
Cleveland	S. S.	Stockton ..	444	,, Certificate rec[d]. Mar. 3.
Nimrod	S.	Cork	351	Lost.
Vistula	S. S.	Londonderry	135	Wrecked.
Elk	S.	Glasgow ..	314	Sunk. Canc[d]. Sep. 17.
Edinburgh	S. S.	Leith	573	Lost.
Tornado	..	Glasgow ..	1075	Abandoned.
Windsor Castle (steel)	S.	Greenock ..	93	Lost.
Barnsley	S. S.	Grimsby ..	340	Near Heligoland.
Hungarian	S. S.	Glasgow ..	487	Off Cape Sable, Nova Scotia.
Bothnia	S. S.	Hull	615	Supposed to have foundered at Sea
Ondine	S. S.	Waterford ..	309	Off Beachy Head. [about Dec.
Pomona	S. S.	,,	995	Off Gozo, in the Mediterranean.
1861.				
Lyra	S.	Londonderry	359	Foundered off Morecambe Bay.
Prince Alfred	S. S.	Leith	430	Off Flamborough Head.

enemy's fire. One hired iron transport in the Baltic, taking home sick, struck once and became a total wreck. If the in-shore squadrons in the Baltic had been iron, nearly all must have been lost unless they had kept out of all the intricate navigation, and left that important service unexecuted.

In the five years since the Russian war the average of wooden ships at sea has been 120; of which the "Raleigh" and "Polyphemus" have been lost. The "Raleigh" might have been saved, could ships and men have been spared from the important service going on. Even allowing three to have been lost since 1854, it would only amount *to 2 per cent. in seven years,* instead of 40 *per cent.* in six years of the iron troop-ships.

The China war lost us the "Transit" and "Assistance," out of the few iron ships we have; while, out of the large number of wooden ships, the "Raleigh" alone was lost, though so many have been on shore. The two iron ships went on rocks in dead smooth water at low speed that would have done no injury whatever to wooden ships had they been on shore.

There are a few smaller iron ships not included; they have been chiefly on the coast of Africa where there is no chance of getting on shore. But as none of our numerous gun-boats at work have been included among the wooden ships; the comparison made is a fair one.

We have about 300 timber ships now in commission, and, if they had been lost at the same rate as iron ships, we should have lost 100 of them in the last three years; instead of which we have not lost a single timber ship by getting on shore. This ought to bring the fullest conviction to all unprejudiced minds.

Iron ships are sinking rapidly in the estimation of all unprejudiced and disinterested persons, and especially underwriters and actuaries of Marine Insurance Companies, whose practical judgment will do more to drive them out of favour altogether than anything that I have said against them. And the proceedings in the House of Commons on the 11th April last, on the resolution moved by Mr. Lindsay, and supported by civil engineers deeply interested as iron-masters, prove that timber ships are rising in estimation, and that the days of timber ships are not over and never will be over.

Another point, intimately connected with the question of armour-plates and iron ships is, the alleged depreciation in the quality of manufactured iron. So far from the quality of iron having been improved in this country, since its production has been enormously increased by smelting it in vast quantities in the iron districts, instead of being produced in small quantities in rural districts, as it was two or three hundred years ago, the deterioration has been in proportion as the quantity produced has increased. It is said that no such iron as that of which the railing round St. Paul's cathedral was made, now exists; and it may fairly and safely be pronounced that quality will be further deteriorated by the shifts to which ironmasters are put to supply the demand.[32]

The iron question is now brought to a sort of settlement by what has been stated in the House of Commons on the navy estimates, viz:—Iron-cased vessels to be provided for home defences, but pronounced unfit for ocean or colonial service, for which timber ships would ever be required. We are now building 10,000 tons of timber ships. The French build their iron-cased vessels of timber, but we persist in building ours of iron; and this is the great distinction between the French and English system, and which is directly the reverse of that upon

---

[32] As bearing on this question, I would refer to an able article on the Iron Manufacture in No. 217 of the ' Quarterly Review,' and more particularly to the Postscript to that article, contained in No. 218, wherein the writer states that the quality of the iron manufacture has deteriorated to so great an extent that a Government Commission has been established to investigate the question. He asserts, that at present the Government possesses no security as to the quality of iron supplied. Contractors being reduced to the lowest possible estimate, are precluded from availing themselves of the best materials ; and, with pig-iron ranging between 45s. and 105s., they are compelled to introduce as much bad with as little good as they can, relying on the intermixture of various qualities for the correction of their respective defects. Allowing that manipulation brings out the quality of all kinds of iron, the writer states that the limit is soon reached beyond which the inferior qualities cease to be improved by it, and that they are rendered absolutely worthless by the processes required to bring the superior quality to perfection.

On the question of armour-plates being manufactured by hammering or rolling, he states his reasons for concluding the latter method to be preferable; and draws the important conclusion, that, whatever kind of manipulation be employed, true economy is secured only by the use of the *best iron*, while he considers that the efficiency of armour-plates would be secured by diminishing their size, not only because they are more cheaply made and more readily repaired, but because, when a plate is struck by a ball, the reaction at the extremities is increased by length.

which the advocates of iron ships urged our imitation of " La Gloire." With respect to that ship it is admitted that she is nothing more than a floating battery on a large scale, fore-and-aft rigged, and whose sailing speed must be very small; that she is so burdened with heavy armour-plates that she cannot go well in a heavy sea; that such ships will do tolerably well in smooth water, but in a heavy sea will be total failures. And these are exactly the points for which I have been always contending with respect to iron-cased vessels.

Firm of purpose, reliant on my scientific and intellectual resources and my long experience, and nothing daunted by the thunders of the press, I vindicate the opinion I have given on the utter unfitness of iron vessels for all purposes of war, which I think no unprejudiced person can now reasonably dispute. That opinion I gave to the Prime Minister of this country twenty years ago. I have been considering it ever since, and abide by it firmly. I believe it is the opinion of all the experimenting officers and committees that have been engaged in testing the question, that iron ships are unfit for battle purposes; and that they are dangerous as transports for troops or emigrants, let the cases of the "Birkenhead," the "Royal Charter," and the vast loss of iron ships, compared with those of wood, answer. This is the opinion of all the naval administrations of this country within the period to which this paper relates. I know it is the opinion of all foreign naval administrations and professional officers, those of France not excepted; and I am sure it is the opinion of every unprejudiced and disinterested man in this country.

CONCLUSION.

To understand rightly the whole subject of nautical and naval science requires the possession of a more extensive range of knowledge than any other *métier*. The philosophy of the nautical and naval profession comprehends, not only an intimate acquaintance with mathematics generally, but also a perfect knowledge of every department of mechanics, theoretical and practical. The naval architect must understand the mode of computing the stability of vessels in water at different angles of heeling. He should be able to calculate the direct, lateral, and vertical resistances of the fluid; and determine the height and breadth of the sails, and the position of the " centre of effort " of the wind upon them, so as to be able to ascertain their moment round the ship's centre of gravity. In steam-vessels, the marine engineer must be able to construct a perfect steam-engine, boilers of a very complex description, and also apply a propelling apparatus of the most appropriate kind, and, finally, combine all these in a perfect and well-proportioned whole. To these may be added the knowledge of mineralogy, metallurgy, and chemistry, as in the present instance.

Undoubtedly the question now in discussion can scarcely be approached by anyone who does not possess some knowledge at least of the subjects alluded to : yet in discussing a question which it is of the very greatest importance to the country should be fully, fairly, truly and soundly discussed and investigated by all the requisite lights of science, we find men rushing out of their depths into the discussion without possessing any of the qualifications that are necessary for a safe settlement of the matter in controversy, as if by miraculous intuition a civilian could improvise mastery of all the branches of naval, military, and general science which it takes other mortals their entire lives' labour and intense study to acquire in their different professional avocations— a presumption of which the article in the 'Quarterly Review' for October contains a curious specimen. Whilst the columns of the periodical and daily press are freely opened to the most crude, ignorant, unscientific, and unfounded assertions, which are received as evidence without being sifted by competent judges, they are too often closed against evidences based on statements of fact, upon which only a right conclusion upon questions of fact can be presented in a state favourable to a correct and exact conclusion. Never since the days of Captain Warner has the public mind been in such a state of hallucination as that to which it is brought by the controversy on iron-sided vessels—as if, on the appearance of a solitary and not successful vessel of that description, we were so frightened out of our propriety as to rush at once into a revolution, not merely on matters of opinion, but in spite of matters of fact, and of far more vital importance to the country.

If I might so far presume as to believe that this 'Postscript to the Naval Gunnery' may be read by a sufficient number of intelligent persons to constitute an influential portion of public opinion, and if I might venture to believe in the "flattering unction" that I am considered to be some authority on those matters, I would earnestly recommend my countrymen to leave such questions as these to the naval administration, with the counsel and advice of such professional men as the Government may think fit to consult; and assure the country that there never was a case which so little needed agitation, and in which there is no cause for alarm, doubt, or difficulty in determining —by a sober, calm, and dispassionate discussion—the matter in question. The existing and preceding naval administrations of this country have for some years had their attention fixed upon the important subject of iron defences; they have neither gone too fast nor too slow; they have neither done too much nor too little; they have taken a middle course in establishing facts and results obtained by the numerous and most valuable experiments which first appeared before the public accurately recorded in Section XII. of the fifth edition of the 'Naval Gunnery.' Unacquainted with those practical results, or, as it appears in the article in the 'Quarterly Review,' ignoring or not consulting the facts of which the writer of that article was in possession, this great question has been treated, distorted, tortured, and embittered [33] in a manner discreditable to some portion of the press of this country.

Having carefully revised the whole of the experiments detailed in Section XII., and studiously reconsidered the deductions at

---

[33] In common with all right-thinking persons, I deem the proper course to pursue with anonymous letters, involving personal accusations, is to commit them to the flames at once; but, to show the animus of ignorant and deluded, perhaps personally-interested parties, advocating the iron-defence question, I cannot forbear making public the following specimen :—

"Sir,   "London.

"Is it—can it be seriously possible that you maintain the astounding proposition that wooden ships are a match for iron-cased ones? I beg you, Sir, to consider well what you advise in this matter. In the event of the certain defeat and disaster which our navy will sustain if your advice is adopted, popular indignation may place your life in jeopardy.

"Sir! you must be mad or incurably prejudiced on this point. It is monstrous that a man should deliberately come to the conclusions you have done, in the face of so much evidence to the contrary.

"AN ENGLISHMAN."

which I had arrived, I adhere firmly to the first conclusion, that vessels formed wholly of iron are utterly unfit for all the purposes and contingencies of war. I ground that opinion upon the incontestible fact, that a plate of wrought-iron of the best quality, 6 feet square and 8 inches thick, leaning upon, but not in contact with, immense slabs of granite by which it was supported, was penetrated, cracked, and broken up by 68-pounder shot at 600 and 400 yards' distance, with a charge of 16 lbs.

Secondly, I maintain that no ship has yet been produced capable of resisting the penetrations and impacts of heavy shot, fulfilling at the same time all the requirements which a sea-going vessel must possess.

Thirdly, I have examined the construction and considered the results of the experiments tried against Mr. Jones's angular iron-plated ship, and have shown that such a vessel would be washed over by a heavy sea, could have no rigging, on account of the deflection of the shot striking her angulated sides, and from the same cause would imperil those she was intended to defend.

Fourthly, I have closely examined, considered, and computed the weight of material in Captain Coles's scheme for remedying the defects in that of Mr. Jones's by placing the armament in revolving towers, and shown its impracticability.

And lastly I assert, on information on which the reader may rely, that "La Gloire" *frégate blindée* is a failure as a sea-going ship—that she is really nothing more than a *batterie flottante* upon a large scale, so burdened with the weight of armament, and loaded with 820 tons of armour-plates, that she is not capable of ocean-service.

Having fully, carefully, and dispassionately considered all these facts and circumstances, I can arrive at no other conclusion than that all the attempts that have been made in France, and by countervailing measures in England, to render iron-cased ships fit for ocean service, either as cruisers or as types for an ocean fleet of iron-cased vessels, have so far proved abortive.

It has been stated in page 15 of this Postscript, that plates capable of resisting the penetration of 68-lb. shot should not be less than $4\frac{1}{2}$ inches thick; but this is, by universal agreement, the maximum thickness that ships can bear to fulfil all the requirements of sea-going vessels.

F

The "idea" of Napoleon III. (preconceived, however, by Paixhans thirty-six years ago), of creating an ocean fleet of iron-cased vessels, fit for active and aggressive services on the sea, having proved to be erroneous by the failure of "La Gloire" as a sea-going ship, it is clear that those vessels can only be used for defensive purposes; and so it appears that these, together with the tortoise angulated gun-boats, of which a great number are forming, he proposes to apply to the protection of the coasts of France, which are everywhere strengthening by the construction of land-batteries.

The remarkable change consequent on the failure of "La Gloire" as a sea-going ship,[34] reverses the case entirely as it concerns us, and relieves us from the terror inspired by the announcement of the complete success of a type on which a fleet of iron-covered ships would be completed in 1861, and leaves us at liberty to think calmly and prudently as to what we should do. This no doubt will be the preparation, by similar means, against any attempt that might be made with such vessels to cross the Channel, though they are not fit to keep the sea, and to continue this defensive measure so long as Napoleon keeps up his iron-cased ships ostensibly for the protection of his coasts and arsenals, which we assuredly will never meditate attacking whilst we remain his faithful ally, nor are we likely under any circumstances to commit any unprovoked aggression on France.

The Emperor will soon find out that in appropriating floating batteries to coast defences, he has fallen out of one error into another, by draining the naval resources and treasury of France to provide for the defence of coasts which it is not our intention

---

[34] While the 1st edition of this 'Postscript' was passing through the press, a confirmation of what I have asserted of "La Gloire" appeared in the 'Times,' extracted from a letter inserted in the semi-official 'Moniteur de la Flotte,' dated Dec. 2nd. There is first an admission that her guns are *not sufficiently out of the water*; and then, alluding to new floating-batteries proposed to be built, it is stated that "They are not intended for going to sea, but solely for the protection of the entrances into ports and rivers. It is chiefly for this service that steel-plated frigates are constructed, as it is not considered safe to send them to sea alone. The general system of the defence of the coasts is to be completed by cutting down the old sailing-ships and plating them. The construction of ships of great speed has naturally directed the attention of the Government to the necessity of adopting greater precaution for the defence of the coasts of Brittany and Normandy. These coasts are now much exposed to the danger of a *coup de main*, which, though not probable, is at least possible."

to attack. As to invasion, thanks to the patriotic spirit which the menace of such an attempt has evoked throughout the country, *that* danger, if not passed, is passing; and, if firm to our purpose of putting England, once and for aye, into a state of security, invasion will never be attempted: even did the portentous state of affairs throughout Europe, produced by Napoleon's Italian war, admit of his concentrating his mind and his forces upon that project.

The system of iron-sided vessels has, so far, produced no practical advantage to either of the parties prosecuting that project; but has impoverished both by the enormous sums of money which it has cost, and it will be an equal relief to both parties when that mania shall have worked itself out. It is well stated, in the leading journal (' Times' of the 31st Oct. which has taken much part in this controversy, that "we are working in the dark,—that the new system of iron-cased ships will lead to hopeless and useless extravagance;" and another journal of high standing (the ' Morning Herald') states that both parties are acting in mistake.[35] I entirely concur in those views. I believe that we are acting against the laws of nature by putting iron to a use for which it is unfit, on account of its specific gravity, which is eight times greater than the weight of an equal bulk of water, and consequently can only be applied to vessels of upwards of 4000 or 5000 tons, and not to vessels of rational and practical magnitude. Prodigies, perhaps, they may be of monster ships; but as profitless, I think, to the defence of the country and the security of our empire as the "Great Eastern" has been to commerce. I think the iron mania, which is subsiding here, will lapse in France; and I should not be surprised that ere long *les frégates blindées* of 1860 will be as much forgotten as *les vaisseaux cuirassés en fer* of 1824 are now.

---

[35] In an article in the ' Morning Herald,' of Nov. 3rd, the writer observes, "If we were called upon to prophesy on the subject, we should anticipate that after a few years (should unhappily a war take place between England and France, or between us and any other iron-coated navy) it will be found that the system is a mistake on both sides; that it will in no degree affect the results of warfare; and will, therefore, have amounted simply to each country having thrown so many millions into the sea."

TABLE OF PRACTICE carried on with Shell, against Iron Plates of various thicknesses, at Shoeburyness.

Date.	Number of Rounds.	Ordnance.	Projectile.	Charge.	Thickness of Plate.	Results.
				lbs.	Inches. Iron.	
22nd Sept. 1859	1	80-pr. gun (Armstrong's)	Cast-iron shell	10	2¾	Shell broke up into splinters.
Do.	2	Do.	Do.	Do.	Do.	Shell broke into pieces, and passed through plate.
Do.	3	Do.	Do.	Do.	3	Shell crushed to pieces.
Do.	4	Do.	Do.	Do.	Do.	Shell broke into pieces.
Do.	5	Do.	Do.	12	Do.	Shell struck and broke on plate.
Do.	6	Do.	Do.	10	Do.	Do. do.
Do.	7	Do.	Do.	Do.	2¾	Shell struck and broke into pieces, passing through plate.
23rd Sept. „	2	68-pr. gun (Service)	Common shell	16	Do.	Shell broke into pieces, chiefly found in front of target.
11th Oct. „	7	Do.	Shell filled with water	16	Steel. 3	Broke plate, shell not reported as broken.
1st March, 1860	5	80-pr. (Armstrong's)	Cast-iron Shell	10	Iron. 2	Shell broke up, piece of plate punched out.
Do.	6	Do.	Do.	Do.	Do.	Shell struck in the same hole as No. 5, chipped next plate, and broke up.
Do.	7	Do.	Do.	Do.	Do.	Shell broke up, and passed through plate.
Do.	8	Do.	Do.	Do.	Do.	Do. do.
Do.	9	68-pr. (Service)	Weighted shell	16	Do.	Shell passed through plate near an old hole, not reported as broken.
Do.	13	Do.	Common shell	Do.	Do.	Shell broke up.

APPENDIX.

It having been asserted that, in my Paper on "Iron Ships and Iron-cased Ships," read at the meeting of the Institution of Naval Architects, held February 28, 1861, I had retracted the opinion I previously held as to iron-built vessels, I think it well to quote the contents of the paper referred to, by way of Appendix to the second edition of the 'Postscript.'

IRON-CASED VESSELS OF WAR.

THE RIGHT HON. SIR J. S. PAKINGTON, Bart., G.C.B., D.C.L., President, after some preliminary remarks, said :—

"In connection with this subject [whether or not armour-covered ships ought to be primarily constructed of wood or of iron] I wish to state, that from an individual to whom the British public are much indebted for the attention he has devoted to the defences of this country—I mean General Sir Howard Douglas—we have just received a communication which I believe it is his wish should be read to this meeting to-day. Sir Howard Douglas is not now in London, owing to ill-health, and therefore he is unable to attend here personally; but he has sent us a paper on the subject. I understand it will not occupy more than a few minutes in reading, and although it has not been put down upon the Proceedings of the day, I think the general feeling of the meeting will be that it will be an act of courtesy to Sir Howard Douglas, and may prove advantageous to the Institution, if that paper is read as the first part of our business to-day. I propose, therefore, that that course be taken, and after Sir Howard's paper is read, we will proceed to hear the other papers that are put down for the consideration of this meeting."

ON IRON SHIPS, AND IRON-CASED SHIPS.

By GEN. SIR H. DOUGLAS, BART., G.C.B., D.C.L., F.R.S., &c.

In my remarks upon the subject of iron ships and iron-cased ships, I have taken care not to confound these two questions with each other. I consider the "Warrior" and the other vessels now being built of timber combined with iron,¶ to belong to the category of iron-cased ships; for although the only timber used in the formation of the "Warrior" consists of two layers of wood, 8 and 10 inches thick respectively, placed behind the plates, yet it must be observed that, but for the timber, by which the plates are backed up, the side of the ship would not be shot-proof, nor could the plates be firmly fixed. Timber being thus indispensable to the formation of iron-cased ships, places them constructively in the category of ships formed of a combination of wood and iron, distinct entirely from ships formed wholly of iron.

In stating the conclusion at which I have arrived, that iron-cased ships are not invulnerable to the penetrations and impacts of heavy solid shot, I do not deny that they are less vulnerable by being so protected than ships that are not so covered, and although we would not have initiated such a system, yet so long as our neighbours—the French—persist in building iron-cased ships, we *must* do so likewise, and that in a manner to keep well ahead of anything the French or any other Power may do for aggressive purposes. I think, therefore, the country is much indebted to Sir John Pakington for having had the moral courage and the administrative enterprize to effect these objects, and that on a scale adequate to satisfy all the requirements which such vessels demand, and which cannot be obtained by vessels of the displacement of "La Gloire."

I have stated in my 'Postscript' a fact well known to ship-builders,—that the bottom of a well-built copper-fastened timber ship scarcely ever wears out, but will at least wear out three tops. In iron ships it is the reverse : one top will wear out three bottoms. This is proved by iron plates now at Lloyds' for the inspection of underwriters. One taken from the bottom of the ship "Jamaica," of 421 tons, built at Glasgow in 1854, worn through and destroyed from bilge-water within and sea-water without; and another taken from the ship "G. F. D.," built at Newcastle in 1849, likewise entirely perished and destroyed. These facts, and the number of iron ships which have lately been wrecked, have produced considerable effect among the Underwriters at Lloyds'; and if many more such cases as the loss of the "Connaught," the "Queen Victoria," and the "Victor Emmanuel" take place, I do not think that insurance will be effected on iron ships against sea risk but at increased premiums, and particularly against war risk in a naval war, such for instance as one with our neighbours, seeing that shot of the most feeble calibre would go through and through them, that they are incapable of carrying armament, either pivot guns or broadside guns adequate to defend themselves—and that such vessels could not safely be employed as troop-ships in war time. Even the "Himalaya," built in 1853, by Mare and Co., Blackwall, successful though she has hitherto been as a troop-ship, when no foe was to be met on the sea, yet the plates in her bottom being only $\frac{7}{8}$ of an inch thick, and tapering off to $\frac{10}{16}$ fore and aft, could not be risked for transporting troops in a real naval war.

Numerous instances occur of wooden ships getting off the strand on which they have struck, without apparent damage. The "Nile," of 1,200 tons, got on shore with 1,200 tons of heavy cargo on board, on the rocks near Vizigapatam, in December, 1853, and laid grinding there for 14 hours. She came off, and proceeded home, however, without making a drop of water; although, when taken into dry dock, it was found that her keel had gone, and many of the bottom planks had rubbed nearly through. Again, in 1854, the "Alfred," with a full cargo, outward bound, met with a similar accident; and when in dock at Calcutta, shewed the same damage as the "Nile," but reached home in safety. H. M. frigate the "Pique," of 40 guns, got on shore, and remained for several hours on the coast of North America; but she got off with the loss of her keel, and reached home in safety. Had these vessels been iron ships, all of them would have been lost. To these cases many others might be added.

With respect to vessels formed wholly of iron, I set out with stating in my 'Naval Gunnery,' pages 133-5, and in the 'Postscript to the Naval Gunnery,' page 6, that in vessels constructed wholly of iron plates $\frac{5}{8}$ of an inch thick, the weight of material in the shell of a ship is considerably less than that of a timber vessel of the same dimensions, and that they will therefore carry a greater weight of cargo and have greater capacity for stowage (on account of the thinness of their shell) than timber ships, and that thus iron may not only be made to float, but to carry a cargo of greater weight and greater bulk than a timber ship of the same dimensions. This theorem, explicitly laid down, keeps me harmless from many of the taunts made in a recently published pamphlet, with a design or not for me—such as that iron vessels sink deeper in the water, on account of the weight of the iron, than timber ships; that iron cannot swim, and cannot be made buoyant as timber, &c.: all of which is a question of displacement, which I have shown may be greater in iron ships than in timber ships of the same dimensions. But, as I have shown in the 'Naval Gunnery,' and in the 'Postscript' thereto, at the pages to which I have already referred, the danger to life and property of those thin-skinned vessels is such as, in my opinion, to render them utterly unfit for the purposes and contingencies of war, and likewise for purposes of commerce in war times.

The reason why the bottoms of the "Warrior" and other iron-cased ships, which are now being built, are not formed of timber, is not a denial of the proofs exhibited of the perishable nature of iron when long exposed to the corroding effects of salt, bilge, and sea water, exhibited in the plates taken from the bottoms of the ships "Jamaica" and "G. F. D.;" but because timber cannot be got of scantling requisite for building ships of such enormous tonnage. But this was not so in the contract for building the iron-cased timber frigate for the service of Russia,* although she was to be of 4,200 tons' displacement, length 300 feet, breadth 55 feet, and total depth 55 feet inside, and was to be covered fore and aft with 4½-inch iron plates, the total weight of which was 1,250 tons, at a cost of 37l. per ton, and her engines 1,000 horse-power; but it was found that, with such a top-weight, the speed could not be more than eleven knots—so true is it that speed and metallic protection throughout are antagonistic. It is computed that to construct a vessel of the description of the "Warrior" in such manner that the speed may be increased from 14 to 18 knots—the speed, it is said, of the new Irish steamers,—the tonnage must be increased from 9,000 to 15,000 tons' displacement, and the power of the engines be trebled.

The numerous losses continually occurring of merchant steamships, nine-tenths of which are formed of thin iron plates,—particularly the loss of the "Connaught," the "Queen Victoria," of 1,152 tons, built at Newcastle in 1854, and of the "Victor Emmanuel," of 520 tons, belonging to Messrs. Joyce and Co., and which broke up as soon as she struck, like a glass bottle against a stone, and the "Metropolis" iron steamship, built in 1857, recently wrecked at Jersey, and which, like the "Royal Charter," broke in two,—have produced great disinclination on the part of underwriters, and determined some to refuse to insure iron ships. In the wreck of timber ships there is always some, and generally a considerable compensation to the underwriters from salvage. What was the salvage on the "Birkenhead," the "Royal Charter," and many iron vessels that have been lost? The salvage on the "Victor Emmanuel," which cost 10,000l., was 20l.

To endeavour to restore confidence in vessels formed wholly of iron, it appears that committees have been formed, and Chambers of Commerce engaged in framing improved regulations for the

---

* See p. 58.

building and strengthening of iron ships. But any such prospective regulations, however useful for the future, will, to a certain extent, cry down the seaworthiness of all iron vessels constructed anterior to the promulgation of the new regulations. But many of the ships recently lost were new ships. The "Connaught" was on her first voyage, the "Queen Victoria," stranded at Plymouth, was a new ship, and the "Victor Emmanuel"—in which all hands, with the exception of a boat's crew, perished—was a new ship.

I may be allowed to feel profoundly the loss of life in all the wrecks of iron ships sacrificed to the manifest dangers to which iron ships are liable. In an awful shipwreck which I suffered in October, 1795, the ship was bilged as soon as she struck, but she held together for more than forty-eight hours, during which time, it is true, some of my brother officers, numbers of the soldiers, half the crew, and all the women and children perished; but the survivors succeeded miraculously in effecting a landing from the bowsprit of the wreck. Had the "Phillis" been an iron ship, however, all hands would have perished in five minutes, as did the troops and the crew of the "Birkenhead," the crew and passengers of the "Royal Charter," and my career would have been cut short ere I had attained my twentieth year. I know something of that description of architecture the foundations of which are laid in Britain's native element—THE SEA. I have, at least, carefully studied the principles upon which the equilibrium and stability of floating bodies depend; and I am far from undervaluing or disparaging that stupendous structure which Mr. Scott Russell has produced in the "Great Eastern;" and, without questioning how far she has been successful in answering the expectations, and fulfilling the purposes for which she was designed, it was surely within my province to examine whether she, the only ship yet constructed wholly of iron, is fit for any purposes of war.

But it appears that I know nothing of the seaworthiness or the stability of iron ships;* that the artilleryman should have confined himself to the artillery question; knows nothing about the rolling of ships, nor of the cause of the weakness of the "Birkenhead" and the "Royal Charter;" and has no practical knowledge or experience of the sea.

Now, the roll of a ship is a matter of the very greatest importance in naval gunnery, and I have made it the subject of elaborate consideration under all the circumstances of the case, namely, the lee roll or the weather roll, whether to fire with the rising motion or

---

* 'Army and Navy Gazette,' Feb. 23rd, 1861.

G

the falling motion of the side, whether to fire when the ship is on
the top of the wave, or the trough of the sea; the gunner considers
the roll of a ship, so far as it effects gunnery, to be seriously or
"awfully" great when, irrespective of gunnery, it is not im-
moderate.   There is no comparison to be made between the roll of
the "Great Eastern," carrying no armament, with that of the
"Ariadne" under different circumstances, but overweighted as all
that class of vessels is with heavy armament.

But, Sir, I am not inexperienced nor unlearned in the sea.   I
am a sailor, every inch of me.   I was born to the sea, nurtured,
tutored, devoted, and destined to the sea.   The army was not the
profession of my choice; but I have not permitted my predilections
for the sea to interfere with my professional avocations in the
profession in which I was placed without my consent; and so it
has come to pass that throughout my life I have been studying,
cultivating, and practising both those professions simultaneously,
as opportunities of combined naval and military service might offer;
and these have been abundant.   I could have earned my bread as a
first-rate practical seaman, when I was a young man; and I have
added something of nautical science, and tactical skill, and practical
gunnery, to my knowledge of practical seamanship.   I have crossed
the sea in every class of ship, from a three-decker to a hired armed
cutter.   The 'Naval Gunnery' was conceived by me in the lower
deck of a line-of-battle ship in action, and worked out till completed
on further experience.

I have revelled in reading the glorious traditions of the past
of the British Navy, and no man can feel more intense interest
than I do, in its PRESENT, at this momentous epoch, and at this
critical period in its history.   But I will not follow others in the
plunge which they take to dive into the future of the British
Navy: not from any dread on my part that the naval power of
England can ever be destroyed by open force, single or coalesced,
if right be done; but from the apprehension that a power which
has withstood a world in arms, defeated maritime coalitions of the
most formidable description, and which formed the only obstacle
that stood in the way between the ambition of Napoleon I. and
universal conquest, might possibly thus be tampered with by a
speculative philosophy, which would prescribe to our descendants
the mode and means of warfare for a remote future, and even
provide them with armour and armament.   Britons will not dege-
nerate; we should be thankful for the past, careful of the present,
and leave the future to our sons.

In reply, therefore, to the question—"Iron or wood?—of which shall our fleets be formed?"—I confidently reply, of neither singly; but by a combination of both to constitute that new description of vessels for special purposes in which the French have taken the lead, but which lead we must take out of their hands by constructing iron-cased ships, which, like theirs, should be formed of timber, that is on wooden bottoms having iron-cased sides—the number and strength of these vessels to be extended accordingly as the circumstances of the case and the perfect security of the country may demand.

With respect to ships formed wholly of iron, I adhere firmly to the opinion I have stated—that they are utterly unfit for any of the purposes of war. The "Great Eastern" belongs to that category, and no one can assert that a vessel that may be perforated through and through by 68-pounder solid shot, is fit for such purposes. Being formed of plates proof against shells, no shells would be fired at her, but solid shot would do the work far more effectually.

The trade of the country could not be carried on by iron vessels formed of thin iron plates, which may be perforated by shot of the most feeble calibre now used in any navy, and incapable of carrying armament adequate to defend themselves, whether on the ends, or on the broadsides (see Report of the Commission to enquire into the capability of arming merchant steamers), and which, as we have seen, are dangerous to life and property, and in a naval war could not be insured against war risks.

This strong point in my exposition of the unfitness of thin-plated iron steamers for purposes of war as merchant vessels, has been evaded by proposing to coat vessels with thicker iron; but this would not do unless it be six inches thick if not backed up with wood, and at least four-and-a-half inches if so backed up.

No real test of the resistance of the iron-cased ships to shot, nor of ships formed of thin plates of iron, will be made till tried in a state of war; and then the very existence of the country, its security, and its trade, would be at stake upon a theory, a speculative experiment untried in war.

It has been said that if the "Warrior" be successful we may bid adieu to timber ships; but it would be quite the reverse. Her success would bid adieu to ships formed of thin plates of iron, because if those ships are not made shot-proof by covering their thin skins with marine layers of timber, and covering these with four-and-a-half inch iron plates, they would not be fit for war purposes, and if so covered, would be unfit for commercial purposes in war, having their tonnage either wholly or greatly absorbed, according to their size, by the weight put upon them.

I regret, exceedingly, that I am unable to attend in person the meeting at which I hope this speech will be read; for speech it is, taken from my dictation, and which I am sure the Institution will receive. I should like much to confront, on their own element, certain distinguished naval officers—the advocates of ships built wholly of iron—who, I perceive, have been invited to attend the meeting, and to combat most respectfully and good-humouredly, but most vigorously with them, upon the whole question of iron defences, and to vindicate all I have written, said, or expressed, on this great subject, by all of which I firmly abide.

Howard Douglas.

The President: " I am sure no gentleman present will regret the course I took the liberty of suggesting—that of commencing our proceedings by the reading of Sir Howard Douglas's interesting paper."